Lecture Notes in Statistics　99

Edited by P. Diggle, S. Fienberg, K. Krickeberg,
I. Olkin, N. Wermuth

C. C. Heyde (Editor)

Branching Processes
Proceedings of the First World Congress

Springer-Verlag
New York Berlin Heidelberg London Paris
Tokyo Hong Kong Barcelona Budapest

C.C. Heyde
Stochastic Analysis Group, CMA
Australian National University
Canberra ACT 0200
Australia

Library of Congress Cataloging-in-Publication Data Available
Printed on acid-free paper.

Camera ready copy provided by the editor.
Printed and bound by Braun-Brumfield, Ann Arbor, MI.
Printed in the United States of America.

9 8 7 6 5 4 3 2 1

ISBN 0-387-97989-1 Springer-Verlag New York Berlin Heidelberg

Contents

Editors Preface

150 Years of Branching Processes

It is now 150 years since statistical work done in Paris on extinction of noble and bourgeois family lines by de Chateauneuf stimulated Bienaymé to formulate what is now usually known as the Galton-Watson branching process model and to discover the mathematical result known as the criticality theorem. However, Bienaymé's work lay fallow and the criticality theorem did not emerge again for nearly 30 years, being rediscovered, not quite accurately, by Galton and Watson (1873-74). From that point the subject began a steady development. The original applications to population modelling were soon augmented by ones in population genetics and later in the physical sciences. More specifically, the process was used to model numbers of individuals carrying a mutant gene, which could be inherited by some of the individuals offspring, and to epidemics of infectious diseases that may be transmitted to healthy individuals. The nuclear chain reaction in reactors and bombs was modelled, and various special cascade phenomena, in particular cosmic rays.

Considerable stimulus to the mathematical development of the subject was provided by the appearance of the influential book, Harris (1963) and this was fueled subsequently by the important books Athreya and Ney (1972) in the USA and Sevastyanov (1971) in the then Soviet Union. More recent. additions to the available list: Jagers (1975), Asmussen and Hering (1983) and Guttorp (1991) have greatly assisted the maintenance of the vitality of the subject.

The First World Conference on Branching Processes was conceived of as an anniversary conference and the papers in these proceedings reflect the broad spectrum of subject matter that was traversed at the Conference. The papers have been grouped, by no means uniquely, into six sections. These reflect both past and future for the subject. The first three sections relate to the three difference types of asymptotic behaviour: supercritical, critical and subcritical (but kept viable by immigration). The second three reflect the variants on the classical model types which have been found useful and have been undergoing lively development.

C.C. Heyde

New York and Canberra

December 1994

References

Asmussen, S. and Hering, H. (1983) Branching Processes. Birkhauser, Boston.

Athreya, K.B. and Ney P.E. (1972) Branching Processes. Springer, Berlin.

Guttorp, P. (1991) Statistical Inference for Branching Processes. Wiley, New York.

Harris, T.E. (1963) Branching Processes. Springer, New York.

Jagers, P. (1975) Branching Processes with Biological Applications. Wiley, London.

Sevastyanov, B.A. (1971) Branching Processes, Nauka, Moscow.

SUPERCRITICAL BRANCHING PROCESSES: A UNIFIED APPROACH

Harry Cohn[*]

University of Melbourne, Australia

ABSTRACT

This paper gives a unified probabilistic approach to the limit theory of finite offspring mean supercritical branching processes. It is shown that the key to the convergence in probability of suitably normed branching processes is a law of large numbers for a triangular array of independent random variables. Criteria for convergence together with characterisations of the norming constants and limiting distributions follow from this property. The models considered include the Galton–Watson and the general age–dependent both in the simple and multitype case as well as in the varying and random environment settings. A martingale derived from a weakly convergent subsequence is essential in the proofs.

Key words Branching, Galton–Watson, varying environment, random environment, age–dependent, muti-type, supercritical, martingale, stochastic monotonicity.

1. Introduction

Let $\{Z_n \; ; n \geq 0\}$ be a sequence of nonnegative random variables such that $P(\lim_{n \to \infty} Z_n = \infty) > 0$, a case which in the branching processes theory is known as supercritical. Our main concern is identifying some constants $\{c_n\}$ such that $\{Z_n/c_n\}$ converges in probability or almost surely to a nondegenerate limit as well as characterising that limit. Write $W_n = Z_n/c_n$. For continuous time models we shall replace n by t where t is supposed to run through the set of non-negative real numbers. For a p-type branching process $\{\mathbf{Z}_n = (Z_n^{(1)}, \ldots, Z_n^{(p)})\}$ we shall consider a similar situation by studying $\{W_n = \sum_{i=1}^{p} c_n^{(i)} Z_n^{(i)}\}$ where $\{c_n^{(i)}\}$ are some constants to be identified. This problem has been the object of intensive study over many years and was surveyed in a number of monographs ([2], [4], [24], [27], etc.). However, no pattern appears discernible from the methods described as each proof seems to be tailored to some particular structure.

---

[*] Department of Statistics, University of Melbourne, Parkville, Victoria 3052, Australia.

Two cases are to be distinguished:

(a) $\{W_n\}$ is a uniformly integrable sequence which corresponds to $E(Z_n) = c_n$ and

(b) $\lim_{n\to\infty} c_n/E(Z_n) = 0$.

The first criterion for such a dichotomy is due to Kesten and Stigum [27] who showed that for a Galton–Watson process with offspring distribution $\{p_k\}$, $\sum_{k=1}^{\infty} p_k \log p_k < \infty$ is necessary and sufficient for (a). When $\{W_n\}$ is uniformly integrable and is a martingale $a.s.$ convergence follows from the martingale convergence theorem. If, as in some continuous parameter or multitype models, $\{W_n\}$ is not a martingale but is uniformly integrable it may still be compared with a certain martingale and this is the way $a.s.$ convergence was proved for multitype Galton–Watson [4], Bellman-Harris [5] and Crump-Mode-Jagers processes [29]. The martingales in question were obtained for each case in a different way.

Indeed, for the Bellman–Harris age–dependent process, the martingale

$$e^{-\alpha t} Z_t \int V(x) dA(x,t)$$

where $V(x) = me^{\alpha x} \int e^{-\alpha u}/[1 - G(x)]dG(u)$, m being the offspring mean, $A(x,t) = Z(x,t)/Z(\infty,t)$ where $Z(x,t)$ is the number of individuals at time t whose age is less than x was used in [5].

For the multitype supercritical Galton–Watson process there is a vector u such and a constant ρ such that

$$(u.Z_n)\rho^{-n}$$

is a martingale. This observation proved instrumental in deriving almost sure convergence in the uniformly integrable case (see [4]).

For the (C–M–J) process the appropiate martingale was identified by Nerman in [29] as

$$Y_t = \sum_{j \in \mathcal{I}_t} e^{-\alpha \sigma_j}$$

where $\mathcal{A}_t$ is the σ-field of the biographies of the T_t individuals born up to t, $\mathcal{I}_t$ is the set of individuals born after t whose mothers were born before t, σ_i is the birth time of the ith individual of $\mathcal{I}_t$ and $\{W_j : j \in \mathcal{I}_t\}$ are independent and identically distributed random variables given $\mathcal{A}_t$.

The first solution to the case of (b) above was given by Seneta [33] who identified $\{c_n\}$ for the simple Galton–Watson process by

$$c_n = -\log f_n^{(-1)}(e^{-s})$$

and proved convergence in distribution of $\{W_n\}$ to a nondegenerate limit by means of Laplace transform and probability generating functions. Here f is the offspring generating function and f_n is the nth iterate of f. Later Heyde [25] found the bounded martingale

$$\{\exp(-c_n^{-1} Z_n)\}$$

thereby proving $a.s.$ convergence. The characterisation of $\{c_n\}$ and the limit distribution of $\{W_n\}$ was carried out by Seneta [34] using analytical methods which rely on functional iterates, Tauberian theorems and regularly varying functions. The "Seneta–Heyde constants" (1) have been extended to the multitype Galton-Watson processes by Hoppe [26] and to Bellman-Harris processe by Schuh [30]. Such extensions involve some topological or measure-theoretic complications. However, more complex models like the simple and multitype Crump-Mode-Jagers [14], [16], multitype varying and random environment [17] and [21] as well as general age-dependent processes [18] do not seem to be amenable to analytical methods and another approach had to be followed instead.

We shall present here a unified approach to supercritical models by showing that for the convergence in probability of $\{W_n\}$ to hold a necessary and sufficient condition is a law of large numbers for a certain sequence of independent random variables. Indeed, consider a sequence of real-valued random variables $\{Y_n\}$ that converges in probability to a limit Y with $P(|Y| \neq \infty)) > 0$. Then if x is a number such that $P(Y \in (x - \epsilon, x + \epsilon)) > 0$ for any $\epsilon > 0$ then there are some $\{x_n\}$ such that $\lim_{n \to \infty} P(Y \in (x - \epsilon, x + \epsilon)|Y_n = x_n) = 1$. This folows easily from

$$P(Y \in (x - \epsilon, x + \epsilon)|Y_n \in (x - \epsilon, x + \epsilon)) = 1. \tag{1}$$

and the mean value theorem (for details see [10]). It turns out that a property of the type of (1) derived in terms of weakly convergent subsequences is also sufficient for convergence in probability (see [10]). This is a rather general situation which extends beyond branching processes and does not require the Markov property. In the branching setup (1) expresses a law of large numbers for sequences of independent random variables.

Instrumental in the poof is the key martingale $\{\xi_n(x)\}$ defined by

$$\xi_n(x) = \lim_{k \to \infty} P(W_{n_k} \leq x|Z_n) \tag{2}$$

where $\{W_{n_k}\}$ is a weakly convergent subsequence. The present account is a streamlined version of the methods used in [6], [10], [14], [17] and [18]. We shall next sketch the basic ideas of the approach.

Let us define $\{c_n\}$ as some numbers satisfying the inequality

$$P(Z_n \leq c_n) \leq \mu < P(Z_n \leq c_n + 1) \tag{3}$$

where $P(\lim_{n \to \infty} Z_n < \infty) < \mu < 1$. Write F_n for the distribution function of Z_n/c_n. As $c_n \to \infty$ and $\lim_{n \to \infty} P(Z_n = x_n) = 0$ for any $x_n > 0$ holds for the models we consider here, (3) implies $\lim_{n \to \infty} F_n(1) = \mu$. Thus $\{F_n\}$ converge in one point which makes $\{c_n\}$ "candidates" for norming constants for some sort of convergence of $\{W_n\}$. The point is to show that convergence in probability and in some instances almost sure convergence are obtainable for this choice of $\{c_n\}$. Indeed, if $\{W_n\}$ converges in probability to a limit W as $n \to \infty$ then

$$\lim_{m \to \infty} P(W_m \leq x | Z_n) = P(W \leq x | Z_n)$$

for any continuity point x of F. Then the above martingale $\{\xi_n(x)\}$ converges $a.s.$ to $1_{\{W \leq x\}}$.

It turned out that the converse is also true and provides a criterion for convergence in probability. Indeed, if $\{\xi_n(x)\}$ defined by (2) has for all x an indicator function for a limit then $\{W_n\}$ converges in probability. Such a limit may be characterised. The key is a law of large numbers, which for example in the case of a varying environment Galton–Watson process, becomes

$$\{W_1^{n)} + \cdots + W_{[c_n]}^{(n)}\} \xrightarrow{P} 1 \tag{4}$$

where $W_1^{(n)}, \ldots, W_k^{(n)}$ are some independent random variables given Z_n. Such a law of large numbers implies that $1_{\{W_n \leq 1\}}$ converges $a.s.$ as $n \to \infty$. But (4) obviously implies

$$\{W_1^{n)} + \cdots + W_{[c_n x]}^{(n)}\} \xrightarrow{P} x$$

for any $x > 0$ and this is equivalent to the $a.s.$ convergence of $1_{\{W_n \leq 1\}}$ for any $x > 0$. It also proves the $a.s.$ convergence of $\{W_n\}$. Besides convergence in probability (or almost sure convergence) this law of large numbers proves instrumental in deriving properties of the norming constants $\{c_n\}$, characterisations of the limit distribution F like tail behaviour, continuity and strict monotonicity, etc.

Let us now mention that in the case when $\{W_n\}$ is uniformly integrable a basic martingale is $\{\eta_t\}$ which we shall define in the continuous time case by

$$\eta_t = \lim_{s \to \infty} E(W_s | \mathcal{F}_t) \tag{5}$$

where $\{\mathcal{F}_t\}$ is a family of increasing σ–fields which generate the σ–field on which the process was defined. This martingale can be obtained from (2) by simple integration over x. Since it has a non degenerate limit, to prove convergence for $\{W_t\}$ one needs to show that $\{\eta_t\}$ and $\{W_t\}$ are asymptotically equivalent. Indeed, it is easy to see that if $\{W_t\}$ is to converges in probability then its limit is the same as that of the martingale $\{\eta_t\}$. An approach based on $\{\eta_n\}$ was used in [14], [18], [21].

2. A key martingale

THEOREM 1   Let $\{Y_t\}$ be a random process with Y_t measurable with respect to $\mathcal{F}_t$ where $\{\mathcal{F}_t\}$ is a family of increasing σ–fields. If there exist a sequence of nonnegative integers $\{t_k\}$ and some numbers $\{x_k\}$ such that

$$\xi_t = \lim_{k \to \infty} P(Y_{t_k} \leq x_k | \mathcal{F}_t) \ a.s.$$

exists for all t, then $\{\xi_t\}$ is a martingale.

The proof is left as an exercise on conditioning.

LEMMA   Suppose that $P(\lim_{n\to\infty} Z_n = \infty) > 0$. Then there exists some constants $\{c_{n_k}\}$ such that $\{Z_{n_k}/c_{n_k}\}$ converges weakly to a limit distribution F with $1 \geq F(\infty) > 1 - P(\lim_{n\to\infty} Z_n = \infty)$.

Proof   Define a sequence of constants $\{c_n\}$ by (3) where $P(\lim_{n\to\infty} Z_n < \infty) < \mu < 1$. It follows necessarily that $\lim_{n\to\infty} c_n = \infty$. By the well-known Helly–Bray theorem there always exists a subsequence $\{Z_{n_k}/c_{n_k}\}$ that converges weakly to a limit distribution F. Two cases may arise:

(i) 1 is a continuity point for F, which leads to $F(1) = \mu$

(ii) 1 is a jump point for F, which leads to $F(1) > \mu$.

Thus $F(\infty) \geq \mu > 1 - P(\lim_{n\to\infty} Z_n = \infty)$ holds in either case completing the proof.

REMARK   The constants $\{c_n\}$ depend on μ while μ is at our disposal in the interval $(1 - P(\lim_{x\to\infty} Z_n = \infty), 1)$. Since $F(\infty) \geq \mu$ we may make $F(\infty)$ as close as desired to 1, but $F(\infty) < 1$ and even $F(\infty) = 1 - F(0)$ may still hold for any choice of $\{c_n\}$.

We may show that Theorem 1 apply to a number of branching models. We start with the simple branching process in varying environment. This is a sequence of nonnegative integer-valued random variables defined inductively by

$$Z_{n+1} = \begin{cases} \sum_{k=1}^{Z_n} X_{n,k} & \text{if } Z_n \geq 1, \\ 0 & \text{if } Z_n = 0. \end{cases}$$

where $\{X_{n,k}; k = 1, 2, \ldots\}$ is, for each n a sequence of independent, identically distributed random variables conditional on Z_n.

THEOREM 2   Suppose that $\{Z_n\}$ is a supercritical branching process in varying environment and $\{c_{n_k}\}$ is a sequence of constants such that $\{Z_{n_k}/c_{n_k}\}$ converges weakly to a limit distribution F with $F(\infty) > 1 - P(\lim_{n\to\infty} Z_n = \infty)$. Then for any continuity point x of F

$$\xi_n(x) = \lim_{k\to\infty} P(Z_{n_k}/c_{n_k} \leq x | Z_n) = \begin{cases} P(W_1^{(n)} + \cdots + W_{Z_n}^{(n)} \leq x | Z_n) & \text{if } Z_n \geq 1, \\ 0 & \text{if } Z_n = 0. \end{cases}$$

where $W_1^{(n)}, \ldots, W_k^{(n)}$ are, independent, identically distributed random variables conditional on Z_n. Notice that this martingale does not require that the weak limit of $\{W_n\}$ is a proper distribution. Indeed, the case when the limit may be infinite with positive probability is included. In fact this was the setup for a irregular branching process with infinite offspring mean considered in [8]. (For a general theory of branching processes with infinite offspring mean see [32].)

For a finite mean Galton-Watson process we get the following theorem.

THEOREM 3   *Suppose that $\{Z_n\}$ is a Galton-Watson process with $1 < m < \infty$. Then there exists a random variable W with $P(W = 0) = q, P(W < \infty) = 1 - q$ where $q = P(\lim_{n\to\infty} Z_n = \infty)$ and some sequences $\{W_1^{(n)}, W_2^{(n)}, \ldots, \}$ of i.i.d. random variables distributed as W and independent of Z_n for $n = 1, 2, \ldots$ such that*

$$\eta_n(x) = P((W_1^{(n)} + \cdots + W_{Z_n}^{(n)})/m^n \le x | Z_n).$$

Let $\{Z_t^\phi\}$ be a (C–M–J) process. The data of the model consists of a random point process ξ on $[0, \infty)$ ruling the reproduction ages of an individual, the life length variable λ and a random characteristic process $\{\phi(t)\}$. Write $\mu(t) = E(\xi(t))$ and assume that there exists a finite positive solution α to the equation

$$\int_0^\infty e^{-\alpha t} \mu(dt) = 1.$$

We get the martingale

$$\xi_t(x) = P(\sum_{j \in \mathcal{I}_t} W_j e^{-\alpha(\sigma_j - t)} / e^{\alpha t} \le x | \mathcal{A}_t)$$

where $\mathcal{A}_t$ is the σ-field of the biographies of the T_t individuals born up to t, $\mathcal{I}_t$ is the set of individuals born after t whose mothers were born before t, σ_i is the birth time of the ith individual of $\mathcal{I}_t$ and $\{W_j : j \in \mathcal{I}_t\}$ are independent and identically distributed random variables given $\mathcal{A}_t$.

Suppose now that $\{Z_n\}$ is a p–type Galton-Watson process with offspring mean matrix $M = (M_{i,j})$ which is assumed to have a maximal eigenvalue $\rho > 1$. Take $\{c_n\}$ such that

$$P(Z_n^{(1)} \le c_n) \le \mu < P(Z_n^{(1)} \le c_n + 1)$$

and write W_j for the weak limit of a subsequence of $\{Z_n^{(1)}/c_n\}$ conditioned on $\mathbf{Z}_0 = j$, and $\{W_{j,k}^{(n)} \ k = 1, 2, \ldots\}$ some i.i.d. copies of W_j given $\mathbf{Z}_n$. Then the corresponding martingale is

$$\xi_n(x) = P(\rho^{-n} \sum_{j=1}^d \sum_{k=1}^{Z_n^{(j)}} W_{k,j}^{(n)} \le x | \mathbf{Z}_n).$$

Notice that even if almost sure convergence is known for some functions of a branching process the above martingale may still provide some extra information on the structure of the process. Consider for example a regular branching process with infinite mean for which there exists some function U and norming constants $\{C_n\}$ such that $\{U(Z_n)/C_n\}$ converges a.s. to a finite limit V, where $\lim_{n\to\infty} C_{n+1}/C_n = \alpha > 1$. In this case it was shown in [7] that

$$\xi_n(x) = P(\max(V_1^{(n)}, \ldots V_{[U^{-1}(C_n)]}^{(n)})/\alpha^n \le x | Z_n)$$

where $V_1^{(n)}, V_2^{(n)}, \ldots$ are i.i.d , independent of $Z_n \ n = 1, \ldots$ and distributed like V. A number of properties for V may be deduced from this martingale (see [7]).

There is another martingale related to $\{\xi_n(x)\}$ which proves useful in situations when $\{W_n\}$ is uniformly integrable. We choose for definiteness to work with the continuous time case.

THEOREM 4   *Suppose that $\{W_t\}$ is a uniformly integrable sequence of random variables defined on a probability space $(\Omega, \mathcal{F}, P)$ and $\{\mathcal{F}_t\}$ a family of increasing σ-fields generating $\mathcal{F}$. Assume that*

$$\eta_t = \lim_{s \to \infty} E(W_s | \mathcal{F}_t)$$

exists for all t. Then $\{\eta_t\}$ is a martingale. A necessary and sufficient condition for the convergence in probability (almost sure) of $\{W_t\}$ is that $\{W_t - \eta_t\}$ converges in probability (almost sure) to 0 as $t \to \infty$.

It is easy to see that $\{\eta_t\}$ is derivable from $\{\xi_t\}$ by taking expectations. It is however simpler to notice the martingale property from the definition of $\{\eta_t\}$.

3. Criteria for convergence in probability and almost sure

Suppose that $\{X_n\}$ is a sequence of real-valued random variables such that
(C_1): $\lim_{k \to \infty} P(X_{n_k} \leq x_k) | X_n)$ a.s. exists any time $\lim_{k \to \infty} P(X_{n_k} \leq x_k)$ exists.
We shall say that x is a point of increase for F if for any $\epsilon > 0$, $F(x + \epsilon) - F(x - \epsilon) > 0$.

THEOREM 5   *Assume that (C_1) holds for $\{X_{n_k}\}$ which converges weakly to a limit distribution F. Write $Q_n(x, A) = \lim_{k \to \infty} P(X_{n_k} \in A | X_n = x)$ for a Borelian set A. Then*

$$\lim_{n \to \infty} Q_n(x_n, (x - \epsilon, x + \epsilon)) = 1 \tag{6}$$

for any point of increase x of F, uniformly with respect to $\{x_n\}$ with $\lim_{n \to \infty} x_n = 1$ is a sufficient condition for convergence in probability of $\{X_n\}$.

For the proof see [10]. Assume now that $\{X_n\}$ is a Markov chain with transition probabilities $\{P_n(x, A)\}$ where $P_n(x, A) = P(X_{n+1} \in A | X_n = x)$. We shall say that $\{X_n\}$ is *stochastically monotone* if $P_n(x, (-\infty, y])$ $n = 1, 2, \ldots$ are non-increasing in x for any fixed y. It is easy to see that the simple and multiple Galton–Watson processes are stochastically monotone. The assumption of stochastic monotonicity may be shown to ensure uniform convergence in (6).

For properties of stochastically monotone Markov chains see [1] and [8].

THEOREM 6   *If, in addition to the assumptions of Theorem 5, $\{X_n\}$ is assumed to be a stochastically monotone Markov chain then $\{X_n\}$ converges almost surely.*

For the proof see [9].

4. A law of large numbers

The following classic result for independent random variables may be used to show that for a number of branching models the limit of the martingale (2) is an indicator function.

THEOREM A *Suppose that for each n, $X_{n,1}, \ldots, X_{n,k_n}$ are independent, identically distributed random variables, $R_n = \sum_{i=1}^{k_n} X_{n,i}$, $\lim_{n \to \infty} k_n = \infty$ and $\lim_{n \to \infty} P(|X_{n,1}| > \epsilon) = 0$ for any $\epsilon > 0$.*

If $\{R_n\}$ converges in distribution to a limit F with bounded support, then there exists a constant c such that $F(x) = 0$ for $x < c$ and $F(x) = 1$ for $x > c$.

Theorem A follows from a result of Khintchine (see e.g. Feller [23], p. 585) relating limit distributions of suitably normed and centered sums of independent random variables to infinitely distributed distributions as well as from a corollary to infinite divisibility (see Feller [23], p.177).

THEOREM 7 *Suppose that $\{Z_n\}$ is a supercritical branching process in a varying environment and $\{c_n\}$ are defined by (3). Then*

$$\{W_1^{n)} + \cdots + W_{[c_n]}^{(n)}\} \xrightarrow{P} 1$$

where $W_1^{(n)}, \ldots, W_{Z_n}^{(n)}$ are the random variables defined in the statement of Theorem 2.

This result holds even if the W's variables are infinite with positive probability. In particular for finite mean Galton–Watson process with offspring mean m we get

$$\{(W_1 + \cdots + W_{[c_n]})/m^n\} \xrightarrow{P} 1.$$

This result was first proved by Seneta [35]. Notice that if $E(W) < \infty$ the above property becomes the Khintchine law of large numbers for i.i.d. sequences.

The following result was proved in [12].

THEOREM 8 *Suppose that $\{X_n\}$ is a sequence of nonnegative, independent and identically distributed random variables, and $\{c_i^{(n)}\}$ are some nonnegative constants with $c_i^{(n)} \leq 1$ for all n and i. Write $S_n = \sum_{i=1}^{n} X_i$, $T_n = \sum_{i=1}^{n} c_i^{(n)} X_i$ and $V_n = \sum_{i=1}^{n} c_i^{(n)}$. Assume that $\{n_k^{-1} V_{n_k}\}$ converges to c as $k \to \infty$ where $\{n_k\}$ is a sequence of integers with $\lim_{k \to \infty} n_k = \infty$, and $\{b_k^{-1} S_{n_k}\}$ converges in probability to m as $k \to \infty$ for some constants $\{b_k\}$ amd m. Then $\{b_k^{-1} T_{n_k}\}$ converges in probability to cm as $k \to \infty$.*

This theorem applied in the setup of a (C–M–J) age–dependent process yields the law of large numbers

$$e^{-\alpha t} \sum_{j=1}^{[c(t)]} e^{-\alpha \sigma_j} W_j \xrightarrow{P} 1$$

4. Almost sure convergence

For the simple Galton–Watson process and the (C–M–J) age dependent process *a.s.* convergence follows from the martingale (2) (see [14]). For the multitype process (see [16]) the approach is similar.

The characterisation of norming constants and limit distribution functions in these *a.s.* convergence theorems is obtained from the following classical result for independent variables.

THEOREM B *Suppose that $X_1, X_2, \ldots$ are independent, identically distributed random variables with a common distribution function F, $S_n = X_1 + \cdots X_n$ and $L(u) = \int_0^u x dF(x)$. In order that there exist constants $\{b_n\}$ such that for any $\epsilon > 0$*

$$\lim_{n \to \infty} P(|b_n^{-1} S_n - 1| > \epsilon) = 0 \tag{7}$$

it is necessary and sufficient that $L(u)$ varies slowly at ∞ (i.e., $\lim_{u \to \infty} L(\lambda u)/L(u) = 1$ for all $\lambda > 0$). In this case there exists some numbers $\{s_n\}$ such that $n L(s_n) = s_n$ and (7) holds with $b_n = nL(s_n)$.

Theorem B is due to Feller ([23] p. 236).

We shall apply Feller's result to (C–M–J) processes in the case when only the first moment of Z_t^ϕ is assumed finite (for details see [14]) .

THEOREM 9 *Let $\{\phi(t)\}$ be a right–continuous random characteristic process with $E(\sup_t e^{-\beta t}\phi(t)) < \infty$ for some β with $0 \le \beta < \alpha$ and $\mu(\infty) < \infty$. Then there exist some constants $\{c(t)\}$ such that $\lim_{t \to \infty} c^{-1}(t) Z_t^\phi = W$ a.s. where W is a non-degenerate random variable. If F is the distribution function of W, then F is continuous on $(0, \infty)$, $L(x) = \int_0^x (1 - F(u))du$ is a slowly varying function and $c(t) \sim e^{\alpha t} L(e^{\alpha t})$.*

5. Convergence in probability

Let $\{\mathbf{Z}_n\} = \{(Z_n^{(1)}, \ldots, Z_n^{(p)})\}$ be a p–type branching process where $Z_n^{(i)}$ stands for the nth generation number of type i particles with $i = 1, \ldots, p$ and

$$Z_{n+r}^{(j)} = \sum_{i=1}^p \sum_{u=1}^{Z_n^{(i)}} Z_{u,i}^{(j)}(n, r) \quad ; j = 1, \ldots, p \tag{8}$$

where $Z_{u,i}^{(j)}(n, r)$ is the number of type j offspring at time $n+r$ of the uth type i particle of the nth generation. The random vectors $\{(Z_{u,i}^{(1)}(n, r), ..., Z_{u,i}^{(p)}(n, r)); u = 1, 2, \ldots\}$ are i.i.d. given $\mathbf{Z}_n$. Consider the mean matrices $\{M_n\} = \{(M_n(i, j))\}$ where $M_n(i, j)$ is the expected number of offspring of type j produced by one particle of type i of the nth generation. Define $^kM^{k-1} = I$ where I is the identity matrix. For $n \ge 1$ it will be seen that if $^1M^n = (^1M^n(i, j)) = M_1 \cdots M_n$, then $^1M^n(i, j) = E(Z_n^{(j)} | \mathbf{Z}_0 = \mathbf{e}_i)$ where $\mathbf{e}_i$ is the p-dimensional vector with 1 in the ith place and 0 elsewhere.

Consider the variables $\{W_n = Z_n^{(j)}/E(Z_n^{(j)})\}$ and the following conditions

C_1 $\lim_{n\to\infty} {}^mM^n(i,j) = \infty$ for $i = 1,\ldots,p$ and $m = 0,1,\ldots.$

C_2 For any m and i

$$\lim_{n\to\infty} {}^mM^n(i,j)/{}^1M^n(1,j) := h(m,i) \tag{9}$$

exists and is finite.

C_3 $\{W_n\}$ is uniformly integrable.

THEOREM 10   *Suppose that* C_1, C_2 *and* C_3 *hold. Then*

(a) $\xi_n(x) = P(\sum_{i=1}^{p} h(n,i)\sum_{u=1}^{Z_n^{(i)}} W_{u,i}(n) \leq x|\mathbf{Z}_n))$

and

(b) $\eta_n = \sum_{i=1}^{p} h(n,i)Z_n^{(i)}$

are martingales, where $\{W_{u,i}^{(j)}(n)\}$ *are some random variables independent of* $\mathbf{Z}_n$ *and distributed according to the weak limit as* $k \to \infty$ *of* $\{Z_{u,i}^{(j)}(n, n_k - n)/E(Z_{u,i}^{(j)}(n, n_k - n))\}.$

Write $c_n = E(Z_n^{(j)}).$

THEOREM 11 $\{h(n,i)\sum_{u=1}^{[c_n]} W_{u,i}(n)\} \xrightarrow{P} K_i$ *as* $n \to \infty$ *for* $i = 1,\ldots p$ *where* $\{K_i\}$ *are some constants is necessary and sufficient for convergence in probablity of* $\{W_n\}.$

A condition of the kind considered above for the varying environment case, unlike the classical Galton–Watson case, does not follow from the structure of the system and is not true in general (see Cohn and Nerman). Thus it needs to be added for convergence in probability to hold.

Let us now consider multitype processes with finite variances. Write $(C_{k,i})$ for the covariance of the reproduction vector of an i type individual of the $(k-1)$th generation. Then C_1, C_2, C_3, C_4 and the following condition

$$\sum_{k=1}^{\infty} \frac{\max_{j,l}(C_{k,j}(l,l)/M_k(1,l))}{\min_l^+({}^1M^k(1,l))} < \infty$$

are sufficient for the L^2–convergence of $\{W_n\}$. Here $\min^+$ refers to the the strictly positive quantities.

A slighty stronger condition may be shown to ensure almost sure convergence for $\{W_n\}$. Indeed, it suffices to notice that *a.s.* convergence follows from $\sum_{n=1}^{\infty} E(W_n - \eta_n)^2 < \infty$ and the Borel–Cantelli lemma. Results on multitype processes in vary ıg and random environment are to be found in [17] and [21]. The general case involves properties of positive deterministic or random matrices (see [19] and [20]).

6. Limiting distribution function

The martingale $\{\xi_n(x)\}$ may be used to derive properties of the limiting distribution function F of the sequence $\{W_n\}$. Indeed, consider the concentration function of a random variable X defined by $p =$

$\sup_x P(X = x)$. Let $\{X_n\}$ be a sequence of independent random variables and $S_n = X_1 + \cdots + X_n$. According to a result by Rogozin [30],

$$\sup_x P(S_n = x) \leq A\Big[\sum_{k=1}^{n}(1 - p_k)^{-1/2}\Big]$$

where A is a universal constant. Since $F(x) = E(\xi_n(x))$ for all n and the number of independent summands in $\xi_n(x)$ tends to infinity as n increases to infinity, we see that $F(x)$ may have jump points in addition to 0 only if the summands in (2) tend rapidly to constants. This is not possible in the case of a Galton– Watson process or a (C–M–J) process as we have random variables whose distributions does not vary with n and therefore continuity for F outside 0 follows (see [11], [14], [16] and [18]). We may use the concentration function to prove continuity for F even when the summands in (2) vary with n as was the case with irregular branching processes with infinite mean in [8] and even without the use of concentration function in the case of varying environment models. Indeed, the concentration function method may not apply to some varying environment models. However if we can prove that F is strictly increasing and dense in $(0, \infty)$ then continuity follows. This is however obtainable by using the law of large numbers (4) (see [10]).

REFERENCES

[1] Aldous, D. (1983) Tail behavior of birth-and-death and stochastically monotone sequences. *Z. Wahrschein-lich. verw. Gebiete*, 62, 375–394.

[2] Asmussen, S. and Hering, H. (1983) **Branching Processes**. Birkhauser, Boston.

[3] Athreya, K.B. (1969) On the supercritical age-dependent branching process. *Ann. Math. Statist.*, 42, 1843–1858.

[4] Athreya, K.B. and Ney, P. (1972) **Branching Processes**. Springer, New York.

[5] Athreya, K.B. and Kaplan, N. (1976) Convergence of age–distribution in the one–dimensional supercritical age dependent branching process. *Ann. Probab.*, 4, 38–50.

[6] Cohn, H. (1977) Almost sure convergence of branching processes. *Z. Wahrscheinlich. verw. Gebiete*, 38, 72–81.

[7] Cohn, H. and Pakes, A.G. (1977) A representation for the limiting variable of a branching process with infinite mean and some related problems. *J. Appl. Prob.*, 15, 225–234.

[8] Cohn, H. and Schuh, H.-J. (1980) On the positivity and the continuity of the limit random variable of an irregular branching process with infinite mean. *J. Appl. Prob.*, 17, 696–703.

[9] Cohn, H. (1981) On the convergence of stochastically monotone sequences of random variables and ap-plications. *J. Appl. Prob.*, 18, 592–605.

[10] Cohn, H. (1982) On a property related to convergence in probability and some applications to branching processes. *Stochastic Processes Appl.* 12, 59-72.

[11] Cohn, H. (1982) Norming constantsfor the finite mean supercritical Bellman–Harris process.
Z. Wahrscheinlich. verw. Gebiete 61, 189–205.

[12] Cohn, H. and Hall, P. (1982) On the limit behaviour of weighted sums of random variables. *Z. Wahrscheinlich. verw. Gebiete* 59, 319–331.

[13] Cohn, H. and Hering, H. (1983) Inhomogeneous Markov branching processes: supercritical case. *Stochastic Processes Appl.*, 14, 178–184.

[14] Cohn, H. (1985) A martingale approach to supercritical (CMJ) branching processes.
Ann. Probab. 13, 1179-1191.

[15] Cohn, H. and Klebaner, F. (1986) Geometric rate of growth in Markov chains with applications to population–size–dependent models with dependent offspring. *Stoch. Anal. Appl.*, 4, 283–307.

[16] Cohn, H. (1989) Multitype finite mean supercritical age–dependent branching processes. *J. Appl. Prob.* 26, 3988–403.

[17] Cohn, H. (1989) On the growth of the multitype supercritical branching process in a random environ-ment. *Ann. Probab.*, 17, 3, 1118–1123, 1989.

[18] Cohn, H. and Jagers P. (1993) General branching processes in varying environment. Accepted for publication in *Ann. Appl. Probab.*

[19] Cohn, H. and Nerman, O. (1990) On products of nonnegative matrices. *Ann. Probab.*, 18, 1806–1815.

[20] Cohn, H., Nerman, O. and Peligrad, M. (1993) Weak ergodicity and products of random matrices. *J. Theor. Prob.* 6, 389–405.

[21] Cohn, H. and Nerman, O. Multitype branching processes in varying and random environment. In preparation.

[22] Doney, R.A. (1972) A limit theorem for a class of supercritical branching processes. *J. Appl. Prob*, 9, 707–724.

[23] Feller, W. (1971) **Introduction to Probability Theory and its Applications.** vol. II, 2nd Ed., New York, Wiley.

[24] Harris, T.E. (1963) **The Theory of Branching Processes.** Springer, NewYork.

[25] Heyde, C.C. (1970) Extension of a result of Seneta for the supercritical Galton–Watson process. *Ann. Math. Statist.*, 41, 739–742.

[26] Hoppe, F. (1976) Supercritical multitype branching process. *Ann. Math. Statist.*, 4, 393–401.

[27] Jagers, P. (1975) **Branching Processes with Biological Applications.** Wiley, New York.

[28] Kesten, H. and Stigum, B. P. (1966) A limit theorem for multidimensional Galton–Watson process. *Ann. Math. Statist.*, 37, 1211–1223.

[29] Nerman. O. (1981) On the convergence of supercritical general (C–M–J) process. *Z. Wahrscheinlich. verw Gebiete*, 57, 365–396.

[30] Rogozin, B.A. (1961) An estimate for concentration function. *Theor. Prob.Appl.*, 6, 96–99.

[31] Schuh, H.-J. (1982) Seneta constants for the supercritical Bellman–Harris process. *Adv. Appl. Prob.*, 14, 732–751.

[32] Schuh, H.-J, Barbour, A. (1977) On the asymptotic behaviour of branching processes with infinite mean. *Adv. Appl. Prob*, 9, 681–723.

[33] Seneta, E. (1968) On recent theorems concerning the supercritical Galton– Watson process. *Ann. Math. Statist.*, 39, 2098–2102.

[34] Seneta, E. (1969) Functional equations and the Galton–Watson process. *Adv. Appl. Prob.*, 1, 1–42. Springer, New York.

[35] Seneta, E. (1974) Characterisation by functional equations of branching process limit laws. In *Statistical Distributions in Scientific Work*, Vol. 3, ed. G.P. Patil, et. al., Reidel, Dordrecht, 294– 254.

On the Statistics of Controlled Branching Processes

Jean-Pierre Dion & Belkheir Essebbar*
Université du Québec à Montréal

ABSTRACT

This work deals with estimation in supercritical controlled branching processes $\{Z_n\}$. The control function φ_n acting on Z_n, at time n, is $\varphi_n(Z_n) = \alpha_n g(Z_n)$ where (α_n) are i.i.d. with values in $\mathbb{N}^+$ and g: $\mathbb{N}^+ \to \mathbb{N}^+$ is a known function such that $g(Z_n) \geq Z_n$. The data set is $\{Z_0 = 1, Z_1, ..., Z_n\}$ and the problem is to estimate the growth rate $\theta = \mu\alpha$ and the criticality parameter $\rho = E \log \mu\alpha_0$, where μ is the offspring mean and $\alpha = E(\alpha_0)$. We show, using martingale theory, that $\hat{\theta} = n^{-1} \Sigma(Z_k / g(Z_{k-1}))$ and $\hat{\rho} = n^{-1} \Sigma \log(Z_k / g(Z_{k-1}))$ are consistent and asymptotically normal and that $\hat{\theta}$ is an asymptotic quasi-likelihood estimator (AQL) of θ. Various extensions are considered : controlled branching in random environments, non i.i.d. (α_n) and the case $\alpha_n \to \alpha$, for which $\tilde{\theta} = \Sigma Z_k / \Sigma g(Z_{k-1})$ is a consistent and an AQL estimator for θ.

Key words: controlled branching processes; statistics; growth rate; criticality parameter; martingale; quasi-likelihood estimator.

1. INTRODUCTION

Controlled branching processes or φ - branching processes were introduced by Sevastyanov & Zubkov (1974) and generalized by Yanev (1975) and others. Although many papers have discussed their probabilistic behaviour, none has, so far, treated the statistical problems arising from this model. It is the purpose of this article to consider estimation in supercritical controlled branching processes. An appropriate background is provided by the texts of Yanev & Yanev (1990, 1991), Dion & Esty (1979) and Heyde (1993).

The multiplicative controlled model that we study is a Markov chain $\{Z_n\}$ defined recursively by $Z_0 \equiv 1$ and

$$Z_{n+1} = \sum_{i=1}^{\alpha_n g(Z_n)} \xi_i(n), \quad n = 0, 1, 2, ...,$$

where

(i) $\xi = \{\xi_i(n), i = 1, 2, ...; n = 0, 1, 2, ...\}$ is a sequence of i.i.d. r.v.'s with values in $\mathbb{N}^+$, the positive integers, and with finite mean μ and variance σ^2.

(ii) $\{\alpha_n, n = 0, 1, ...\}$ is a sequence of i.i.d. r.v.'s, with values in $\mathbb{N}^+$ and g: $\mathbb{N}^+ \to \mathbb{N}^+$ is a known function such that $g(n) \geq n, \forall n = 1, 2, ...$. We assume throughout that Var $(\alpha_0) = b^2 < \infty$ and Var $\log \alpha_0 = \eta^2 < \infty$.

* Postal address: Professor J.P. Dion, Département de mathématiques, Université du Québec à Montréal, C.P. 8888, succursale Centre-ville, Montréal, Québec, H3C 3P8.

and

(iii) $\xi = \{\xi_i(n)\}$ and $\{\alpha_n\}$ are two independent sequences.

Then $\{Z_n\}$ is a supercritical branching process with offspring sizes given by $\{\xi_i(n)\}$ and control function defined by $\{\varphi_n(i) = \alpha_n g(i)\}$.

Clearly, if $\varphi_n(i) \equiv i \; \forall n$, it is the Bienaymé - Galton - Watson (BGW) process, while if $g(i) \equiv i$, it is a branching process with random environment (BPRE) (see Yanev & Yanev, 1991). Our multiplication control model is a particular case of the controlled branching process with random environment discussed in Yanev & Yanev (1990).

The data set is $\{Z_0 = 1, Z_1, \ldots Z_n\}$ and the problem is to estimate the growth rate $\theta = \mu\alpha$ and the criticality parameter $\rho = E \log \mu\alpha_0$, where $\alpha = E(\alpha_0)$.

We will show, using martingale theory, that

$$\hat{\theta} = n^{-1} \Sigma(Z_k / g(Z_{k-1}))$$

and

$$\hat{\rho} = n^{-1} \Sigma \log(Z_k / g(Z_{k-1}))$$

are consistent and asymptotically normal and that $\hat{\theta}$ is an asymptotic quasi-likelihood estimator (AQL) of θ. This is in complete analogy with the results of Dion & Esty (1979) and Heyde (1993), on BPRE.

Consistent estimators are also provided for the asymptotic variances. Finally, we will consider some variants of our model, in particular the case $\alpha_n \to \alpha = E(\alpha_k) \; \forall k$, for which

$$\tilde{\theta} = \Sigma Z_k / \Sigma g(Z_{k-1})$$

is a consistent and an AQL estimator for θ.

2.   ESTIMATION OF $\theta = \mu\alpha$

Consider the sample $\{Z_0, Z_1, \ldots, Z_n\}$ and the filtration $(F_n = \sigma(Z_0, \ldots, Z_n, \alpha_0, \ldots, \alpha_n))$.

Since

$$E\left(\left(\frac{Z_{k+1}}{g(Z_k)} - \mu\alpha_k\right) | F_k\right) = 0$$

forms the zero-mean martingale

$$M_n = \sum_1^n \left(\frac{Z_k}{g(Z_{k-1})} - \mu\alpha_{k-1}\right), \; n \geq 1$$

with respect to the filtration (F_n).

One then has the basic result:

Lemma 1 : $M_n \to X$ a.s. and in L_2, for some a.s. finite r.v. X.

Proof : Associate to the multiplicative controlled branching process $\{Z_n\}$ a simple BGW process $\{Z_n^1\}$ using the same sequence of offspring sizes $\{\xi_i(n)\}$. Then

$$g(Z_n) \geq Z_n \geq Z_n^1, \; \forall \; n.$$

Hence

$$E\left(M_n^2\right) = \sigma^2 \sum_1^n E\left(\alpha_{k-1}/g(Z_{k-1})\right) = \sigma^2 \alpha \sum_1^n E\left(\left[g(Z_{k-1})\right]^{-1}\right)$$

$$\leq \sigma^2 \alpha \sum_1^\infty E\left[\left(Z_{k-1}^1\right)^{-1}\right] < \infty$$

by a result of Badalbaev & Mukhitdinov (1990, p.17) on BGW processes. We have shown that $\sup_n E\left(M_n^2\right) < \infty$, and thus we have L_2 convergence of the martingale M_n. Being a zero-mean square integrable martingale, M_n converges a.s. to an a.s. finite random variable on the set $\left[\langle M\rangle_\infty < \infty\right]$, where

$$\langle M\rangle_\infty = \lim_n \langle M\rangle_n = \lim_n \sum_1^n \left(\sigma^2 \alpha_{k-1}/g(Z_{k-1})\right).$$

This last expression is a.s. finite since its expectation is finite; this completes the proof.

An immediate consequence is:

Corollary 1 :

    (i) $(Z_n / g(Z_{n-1})) - \mu\alpha_{n-1} \to 0$ a.s.

    (ii) $Z_n / \mu\alpha_{n-1} g(Z_{n-1}) \to 1$ a.s.

    (iii) $\forall\, \varepsilon > 0,\ M_n / n^\varepsilon \to 0$ a.s.

The main result of this section is the following :

Theorem 1 :

    Let $\theta = \mu\alpha$ and $\hat\theta = n^{-1}\sum_1^n Z_k/g(Z_{k-1})$.

Then, as $n \to \infty$,

    (i) $\hat\theta \to \theta$ a.s.

and

    (ii) $\sqrt{n}\left(\hat\theta - \theta\right) \xrightarrow{d} N\left(0, b^2\right).$

Proof. Let $\bar\alpha_n = n^{-1}\sum_1^n \alpha_{k-1}$. By the SLLN, $\bar\alpha_n \to \alpha$ a.s. Since $M_n / n = \hat\theta - \mu\bar\alpha_n$, part (i) follows readily from Corollary 1 (iii).

Similarly,

$$\sqrt{n}\left(\hat\theta - \theta\right) = \sqrt{n}\left(\frac{M_n}{n} + \mu\bar\alpha_n - \theta\right) = \frac{M_n}{\sqrt{n}} + \mu\sqrt{n}(\bar\alpha_n - \alpha).$$

Using again Corollary 1 and the CLT for (α_k), one obtains part (ii).

In order to provide an approximate confidence interval for θ, one may use the consistent estimator

$$\hat{b}^2 = n^{-1}\sum_1^n \left(\frac{Z_k}{g(Z_{k-1})} - \hat\theta\right)^2$$

for the (presumably) unknown variance b^2.

Theorem 2 : $\hat{b}^2 \to b^2$ a.s. as $n \to \infty$.

Proof. Adding and substracting θ,

$$\hat{b}^2 = n^{-1} \sum \left(\frac{Z_k}{g(Z_{k-1})} - \theta \right)^2 - \left(\hat{\theta} - \theta \right)^2.$$

Since $\hat{\theta} - \theta \to 0$ a.s., it is sufficient to prove that $n^{-1} \sum \left(\frac{Z_k}{g(Z_{k-1})} - \theta \right)^2 \to b^2$ a.s.

By expanding

$$n^{-1} \sum \left(\frac{Z_k}{g(Z_{k-1})} - \mu\alpha_{k-1} + \mu\alpha_{k-1} - \theta \right)^2$$

the arguments given in Dion & Esty (1979, p. 682) apply *mutatis mutandis* and lead to the required conclusion.

3.   ESTIMATION OF $\rho = E \log \mu\alpha_0$

The natural martingale for our model is not M_n but

$$W_n = \prod_1^n \left(Z_k / \mu\alpha_{k-1} g(Z_{k-1}) \right), \ n \geq 1.$$

The martingale property is obvious since

$$E\left(W_{n+1} \mid F_n \right) = W_n E\left(Z_{n+1} / \mu\alpha_n g(Z_n) \mid F_n \right) = W_n.$$

When our model reduces to the BGW case, W_n reduces to Z_n / μ^n and when $g(i) \equiv i$, (a BPRE process), W_n reduces to $Z_n \Big/ \prod_1^n (\mu\alpha_{k-1})$.

Since W_n is a non-negative martingale it converges a.s. to an a.s. finite r.v. W. The L_2 convergence of W_n to W follows from the next lemma and, since $1 = E(W_n) \to E(W)$, ensures that W is not degenerate at 0.

Lemma 2 : For the martingale $\{ W_n, F_n \}$, one has $\sup_n E\left(W_n^2 \right) < \infty$ and hence $W_n \to W$ in L_2.

Proof : We have

$$E\left(W_{n+1}^2 \mid F_n \right) = W_n^2 E\left[\left(Z_{n+1}^2 / (\mu\alpha_n g(Z_n))^2 \right) \mid F_n \right].$$

and as

$$E\left(Z_{n+1}^2 \mid F_n \right) = \sigma^2 \alpha_n g(Z_n) + \mu^2 \alpha_n^2 (g(Z_n))^2,$$

$$E\left(W_{n+1}^2 \mid F_n \right) = W_n^2 \left[1 + \sigma^2 / \mu^2 \alpha_n g(Z_n) \right] \leq W_n^2 \left[1 + \frac{K}{g(Z_n)} \right]$$

where $K = \sigma^2 / \mu^2$. (Recall that $\alpha_n \geq 1$.)

Thus $E W_n^2 \leq E(W_1^2) \prod_1^n \left(1 + KE\left[(g(Z_{n-k}))^{-1}\right]\right)$

and

$$\sup_n E \ W_n^2 \leq E(W_1^2) \prod_1^\infty \left(1 + KE\left[(g(Z_k))^{-1}\right]\right) < \infty$$

since

$$\prod_1^\infty \left(1 + KE\left[(g(Z_k))^{-1}\right]\right) < \infty \Leftrightarrow \sum_1^\infty E\left(g(Z_k)\right)^{-1} < \infty.$$

As a consequence W is not degenerate at 0; in fact $W > 0$ a.s. as is shown in the next Lemma and we will use later the fact that log W is finite a.s.

Lemma 3 : Let $W = \lim$ a.s. W_n, where $W_n = \prod_1^n (Z_k / \mu \alpha_{k-1} g(Z_{k-1}))$, $n \geq 1$. Then $P(W > 0) = 1$.

Proof : It is sufficient to prove that $\left\{W_n^{-1}\right\}$ converges a.s. to some a.s. finite r.v.

Since $\{W_n, F_n\}$ is a positive martingale, $\left\{W_n^{-1}, F_n\right\}$ is a submartingale, with $E(W_1^{-1}) \leq \mu < \infty$. The almost sure convergence of $\left\{W_n^{-1}\right\}$ to some a.s. finite r.v. will follow if we establish that $\sup_n E\left(W_n^{-1}\right) < \infty$.

$$E\left(W_n^{-1}|F_{n-1}\right) = W_{n-1}^{-1}\mu \alpha_{n-1} g(Z_{n-1}) E\left(Z_n^{-1}|F_{n-1}\right).$$

Using a Taylor expansion at $\mu \alpha_{n-1} g(Z_{n-1})$,

$$E\left(Z_n^{-1}|F_{n-1}\right) = \frac{1}{\mu \alpha_{n-1} g(Z_{n-1})} + E\left\{\frac{\left(Z_n - \mu \alpha_{n-1} g(Z_{n-1})\right)^2}{H_n^3}\bigg|F_{n-1}\right\}$$

where

$$H_n \geq \min\left\{Z_n, \mu \alpha_{n-1} g(Z_{n-1})\right\} \geq \alpha_{n-1} g(Z_{n-1}) \quad \text{a.s.}$$

Hence

$$E(Z_n^{-1}|F_{n-1}) \leq \frac{1}{\mu \alpha_{n-1} g(Z_{n-1})} + \frac{\text{var}\left(Z_n|F_{n-1}\right)}{\left[\alpha_{n-1} g(Z_{n-1})\right]^3},$$

where

$$\text{var}(Z_n|F_{n-1}) = \sigma^2 \alpha_{n-1} g(Z_{n-1}).$$

Finally,

$$E(W_n^{-1}|F_{n-1}) \leq W_{n-1}^{-1}\left\{1 + \frac{\sigma^2 \mu}{\alpha_{n-1} g(Z_{n-1})}\right\} \quad \text{a.s.}$$

and

$$E(W_n^{-1}) \leq E(W_1^{-1}) \prod_1^n \left(1 + \sigma^2 \mu E\left[g(Z_{n-k})^{-1}\right]\right).$$

As in the proof of Lemma 2, this inequality implies that

$$\sup_n E\left(W_n^{-1}\right) < \infty \text{ and thus } W > 0 \text{ a.s.}$$

Theorem 3 :

Let $\rho = E \log \mu \alpha_0$ and $\hat{\rho} = n^{-1} \sum \log\left(Z_k / g(Z_{k-1})\right)$.

Then, as $n \to \infty$,

    (i) $\hat{\rho} \to \rho$ a.s.

and

    (ii) $\sqrt{n}(\hat{\rho} - \rho) \xrightarrow{d} N(0, \eta^2)$, where $\eta^2 = \text{Var} \log \alpha_0$.

Proof.

Put $H_n = \log W_n = \sum_1^n \left[\log\left(Z_k / g(Z_{k-1})\right) - \log \mu \alpha_{k-1}\right]$.

Since $H_n \to \log W$ a.s. and $\log W$ is finite a.s., $H_n/n^\varepsilon \to 0$ a.s., $\forall \ \varepsilon > 0$. In particular, $H_n / n \to 0$ a.s. $\Rightarrow \hat{\rho} - n^{-1} \Sigma \log \mu \alpha_{k-1} \to 0$ a.s., or equivalently, from the SLLN for $(\log \mu \alpha_k)$, $\hat{\rho} - \rho \to 0$ a.s., proving part (i).

Taking $H_n/\sqrt{n} \to 0$ a.s. and invoking the CLT for $\{\log \mu \alpha_k\}$, one has

$$\sqrt{n}(\hat{\rho} - \rho) \xrightarrow{d} N(0, \eta^2),$$

where $\eta^2 = \text{Var} \log \mu \alpha_0 = \text{Var} \log \alpha_0$.

Note that Becker's estimator for ρ

$$\tilde{\rho} = \frac{\log Z_n}{n}$$

which is appropriate in a BPRE process (see also Dion & Esty (1979)) is not consistent for our model, unless $g(i) \equiv i$, in which case $\tilde{\rho} = \hat{\rho}$.

4. ASYMPTOTIC QUASI-LIKELIHOOD ESTIMATORS

Using the notation of Heyde's (1993) review paper on estimating equations, let us consider a class of estimating functions for $\theta = \mu \alpha$:

$$\mathcal{H} = \left\{ h : h = \sum_1^n a_k(\theta) h_k(\theta) \right\}$$

where $a_k(\theta) = a_k(\theta; Z_0, \ldots, Z_{k-1})$ is predictable with respect to $(F_n = \sigma(Z_0, \ldots, Z_n))$ and

$$E\left(h_k(\theta) \mid F_{k-1}\right) = 0$$

so that each $h \in \mathcal{H}$ is a zero-mean martingale.

It is well known that within $\mathcal{H}$, the optimal weights $a_k(\theta)$ are given by

$$a_k^*(\theta) = \left[E\left(\dot{h}_k(\theta) \mid F_{k-1}\right)\right]\left[E\left(h_k^2(\theta) \mid F_{k-1}\right)\right]^{-1}$$

$\dot{h}$ denoting the derivative with respect to θ. For our model and data set, an appropriate function $h_k(\theta)$ is

$$h_k(\theta) = Z_k - E(Z_k \mid F_{k-1}) = Z_k - \theta\, g(Z_{k-1}).$$

Hence the optimal (quasi likelihood) estimator is the solution to

$$h^* = \sum_1^n \frac{g(Z_{k-1})}{\mathrm{var}(Z_k \mid F_{k-1})}(Z_k - \theta\, g(Z_{k-1})) = 0.$$

Since

$$\mathrm{var}(Z_k \mid F_{k-1}) = \alpha\sigma^2\, g(Z_{k-1}) + \mu^2 b^2\big(g(Z_{k-1})\big)^2,$$

the equation

$$h^* = \sum \frac{(Z_k - \theta g(Z_{k-1}))}{\alpha\sigma^2 + \mu^2 b^2 g(Z_{k-1})} = 0$$

is asymptotically equivalent to

$$h_1^* = \sum_1^n \left(\frac{Z_k}{g(Z_{k-1})} - \theta \right) = 0,$$

whose solution is precisely $\hat\theta = n^{-1}\sum_1^n Z_k / g(Z_{k-1})$.

In that sense our estimator is also AQL estimator and will provide the shortest confidence interval for θ.

Suppose now the data set is $\{Z_0, \ldots, Z_n, \alpha_0, \ldots, \alpha_n\}$ and let us concentrate on estimating μ. Consider then another filtration $(F_n = \sigma(Z_0, \ldots, Z_n, \alpha_0, \ldots, \alpha_n))$ and a class of estimating functions

$$G = \{ g : g = \sum_1^n a_k(\mu, Z_0, \ldots, Z_{k-1}, \alpha_0, \ldots, \alpha_{k-1})(Z_k - \mu\alpha_{k-1} g(Z_{k-1})) = 0$$

a_k predictable. Then, the quasi-score estimating equation

$$g^* = \sum_1^n \frac{(Z_k - \mu\alpha_{k-1} g(Z_{k-1}))}{\sigma^2} = 0$$

gives the quasi-likelihood estimator

$$\hat\mu = \sum_1^n Z_k \Big/ \sum_1^n \alpha_{k-1} g(Z_{k-1}),$$

the analog of Harris estimator for the offspring mean.

Since $g^* / \langle g \rangle_n = \hat\mu - \mu$, consistency of $\hat\mu$ follows easily from SLLN for this zero mean square integrable martingale g^*.

This result carries over to non i.i.d. (α_k), for instance (α_k) such that $\alpha = E(\alpha_k)$, $\forall\, k$, and $\alpha_k \to \alpha$ a.s.. In that case by the Toeplitz lemma,

$$\sum_1^n \alpha_{k-1} g(Z_{k-1}) \Big/ \sum_1^n g(Z_{k-1}) \to \alpha$$

and $\sum Z_k / \sum g(Z_{k-1})$ is a consistent estimator for $\theta = \mu\alpha$.

5. VARIANTS AND EXTENSIONS.

Theorems 1 - 3 were proved for (α_k) i.i.d. with values in $\mathbb{N}^+$. They remain valid when the (α_k) are i.i.d. with values in $[1, \infty)$ for the model $Z_0 = 1$ and $Z_{n+1} = \sum_{i=1}^{[\alpha_n g(Z_n)]} \xi_i(n)$, where $[x]$ is the integer part of x, $x \in \mathbb{R}$. Details of this assertion have been given in Essebbar (1993, unpublished, Ph.D. Thesis) for the case $g(n) \equiv n$ and they apply as well to general g.

A second extension is to non i.i.d. (α_k). Clearly Theorems 1 and 2 remain valid whenever LLN and CLT hold for the α_k's, while Theorem 3 requires these to hold for $\{\log \mu\alpha_k\}$.

Finally, as our model is a particular controlled branching process with random environment, some of our results can presumably be extended to Yanev & Yanev's (1990) model.

Acknowledgements

We are thankful to René Ferland for stimulating discussions on the topic.

RÉFÉRENCES

ATHREYA, K. and NEY, P. (1972). *Branching Processes*. Springer, Berlin.

BADALBAEV, I.S. and MUKHITDINOV, A. (1990). *Statistical Problems in Multitype Branching Processes*. Fan, Tashkent (in Russian).

BECKER, N. (1977). Estimation for discrete time branching processes with application to epidemics. *Biometrics*, **33**, 515-522.

DACUNHA-CASTELLE, D. and DUFLO, M. (1983). *Probabilités et Statistiques*. Tome 2, Masson, Paris.

DION, J.-P. and ESTY, W.W. (1979). Estimation problems in branching process with random environments. *Ann. Statist.* **7**, 3, 680-685.

ESSEBBAR, B. (1993). *Inférence dans les processus stochastiques discrets*. Ph. D. thesis, Dept. Math. & Info., UQAM, Canada.

HALL, P. and HEYDE, C.C. (1980). *Martingale Limit Theory and its Application*. Academic Press, New York.

HEYDE, C.C. (1993). Quasi-likelihood and general theory of inference for stochastic processes. *Lect. Notes*, 7th Int. Summer School on Prob. Theo. & Math. Stat. (Varna, 1991). Ed. A.Obretenov & V.T. Stefanov, Sci. Cult. Tech. Publ., Singapore, 122-152.

SEVASTYANOV, B.A. and ZUBKOV, A.M. (1974). Controlled branching processes. *Theor. Probab. Appl.* XIX, **1**, 15-25.

YANEV, N.M. (1975). Conditions of extinction of φ-branching processes with random φ. *Theor. Probab. Appl.* XX, 2, 433-440.

YANEV, N.M. and YANEV, G.P. (1990). Extinction of controlled branching processes in random environments. *Math. Balkanica*, **4**, 368-380.

YANEV, N.M. and YANEV, G.P. (1991). Branching processes with multiplication : the supercritical case. *C.R.Acad. Bulg. Sci.*, **44**, 4, 15-18.

ASPECTS OF THE CRITICAL CASE OF A GENERALISED GALTON-WATSON BRANCHING PROCESS

M. P. QUINE,[*]University of Sydney
W. SZCZOTKA,[†]Wrocław University

Abstract

We define a random process $\mathcal{X} = \{X_n,\ n = 0, 1, 2, \cdots\}$ in terms of successive cumulative sums of a sequence $\{K_n,\ n = 1, 2, \cdots\}$ of integer-valued random variables. $\mathcal{X}$ contains as a special case the Galton-Watson process. We discuss some basic results concerning the possible asymptotic behaviour of $\{X_n\}$ as $n \to \infty$ and examine some examples in the case when $n^{-1}\sum_{i=1}^{n} K_i \to 1$ a.s., which corresponds to the "critical" case for Galton-Watson processes.

Key words: Branching process, imbedded random walk.

1   Introduction

Let $K_1, K_2, \cdots$ be integer-valued random variables (rv's) defined on the same probability space $(\Omega, \mathcal{F}, P)$ and define sequences $\mathcal{X} = \{X_n, n = 0, 1, 2, \cdots\}$ and $\mathcal{T} = \{T_n, n = 0, 1, 2, \cdots\}$ by $X_0 = 1$, $X_1 = K_1$, $T_n = \sum_{j=0}^{n} X_j$, $n = 0, 1, 2 \cdots$ and

$$X_{n+1} = \left(\sum_{j=T_{n-1}+1}^{T_{n-1}+X_n} K_j \right) I(X_n \geq 1),\ n = 1, 2, \cdots, \qquad (1)$$

where $I(A)$ denotes the indicator of A.

It is not difficult to check that if $K_1, K_2, \cdots$ are independent and identically distributed (iid) and non-negative then $\mathcal{X}$ is a Galton-Watson branching process and $\mathcal{T}$ the corresponding total progeny process[1]. So (1) allows us to extend the definition of the branching process to the case where the offspring distribution has support on all the integers. In that case, i.e. when $K_1, K_2, \cdots$ are iid with $P(K_1 < 0) > 0$ then $\mathcal{X}$ can be given a branching process interpretation as follows: suppose there are two types of particle and let X_n denote the excess of type I over type II particles at the n-th generation. If $X_n \leq 0$ then $X_{n+1} = X_{n+2} = \cdots = 0$. If $X_n = k > 0$ then the excess of type I over type II particles in

---

[*]School of Mathematics and Statistics, University of Sydney, NSW 2006, Australia.

[†]Mathematical Institute, Wrocław University, Pl. Grunwaldzki 2/4, 50-384 Wrocław, Poland.

[1]Sankaranarayanan (1989, page 3) erroneously claims that $X_{n+1} = \sum_{j=1}^{X_n} K_j$ gives a branching process.

the $n + 1$-th generation is the sum of k iid rv's with the distribution of K_1. In the case where $K_1, K_2, \cdots$ are independent and non-negative but not iid then $\mathcal{X}$ can be regarded as a branching process with varying offspring distributions.

For the process $\mathcal{T}$, on the event $\cap_{n \geq 1}\{X_n > 0\}$ the indicator function in (1) equals unity for $n = 1, 2, \cdots$, so

$$T_{n+1} = 1 + \sum_{j=1}^{T_n} K_j, \ n = 1, 2, \cdots \tag{2}$$

and

$$X_{n+1} = T_{n+1} - T_n = 1 + \sum_{j=1}^{T_n} \tilde{K}_j, \ n = 1, 2, \cdots, \tag{3}$$

where $\tilde{K}_j = K_j - 1$. It follows that on the event $\{T_n \to \infty\}$, $\mathcal{X}$ is imbedded in the random walk $\tilde{S} = \{\tilde{S}_n, n = 0, 1, 2, \cdots\}$, where $\tilde{S}_0 = 0$ and $\tilde{S}_n = \sum_{j=1}^{n} \tilde{K}_j, \ n = 1, 2, \cdots$, in the sense that

$$X_{n+1} = 1 + \tilde{S}_{T_n}, \ n \geq 1.$$

This representation of $\mathcal{X}$ as an imbedded random walk allows us in particular to investigate conditions under which the following "extinction/explosion" results hold as $n \to \infty$:

$$P(X_n \to \infty \text{ or } X_n \to 0) = 1, \tag{4}$$

$$P(X_n \to 0) < 1, \tag{5}$$

and

$$P(X_n \to 0) = 1. \tag{6}$$

Theorem 1 *Suppose $K_1, K_2, \cdots$ are such that, as $n \to \infty$, $\frac{\tilde{S}_n}{n} \to \lambda$ a.s. for some $\lambda \in [-\infty, \infty]$. Then*

(a) If $\lambda \neq 0$, then (4) holds.

(b) If $\lambda > 0$ and $P\left(\tilde{S}_n \geq 0, \ n = 1, 2, \cdots\right) > 0$, then (5) holds.

(c) If $\lambda < 0$, then (6) holds.

It is easy to check that Theorem 1 covers the non-critical parts of the Criticality Theorem for branching processes. In particular, *(b)* relates to any supercritical $(1 < EK_1 \leq \infty)$ branching process; it is a standard random walk result that $P\left(\tilde{S}_n \geq 0, \ n = 1, 2, \cdots\right) = P(V = \infty) > 0$, where $V = \inf\{n : \tilde{S}_n < 0\}$, whenever $E\tilde{K}_1^- < E\tilde{K}_1^+ \leq \infty$ (see e.g. Feller (1971, pages 396-397)).

The situation is less clear when

$$\frac{\tilde{S}_n}{n} \to 0 \text{ a.s.} \tag{7}$$

That is, the "critical" $(EK_1 = 1)$ branching process results cannot be extended so widely. About the most general result we have been able to obtain for $\lambda = 0$ in this case is as follows:

Theorem 2 *If $K_1, K_2, \cdots$ are iid with $P(K_1 = 1) < 1$ and $E(K_1) = 1$, then a sufficient condition for (6) to hold is that there exists a number $C < \infty$ such that*

$$-E(K_1 + \cdots + K_j | K_1 + \cdots + K_j \le 0) \le C \text{ for all } j. \tag{8}$$

Theorems 1 and 2 are proved in Quine and Szczotka (1994). In the sequel, we first consider two examples where (7) holds for non-iid rv's, but (6) and/or (4) fails. Then we give two examples involving iid rv's where (7) and (6) both hold and give a more general iid result. Finally we give a non-iid extension of Theorem 2.

2 The critical case

2.1 Non-iid examples

We first give two examples of what can happen when (7) holds in the non-iid case.

Example 1 Suppose $K_1, K_2, \cdots$ are independent with $P(K_j = 1) = 1 - 1/j^2$ and $P(K_j = j) = 1/j^2$. Then $E(K_j) = 1 + 1/j - 1/j^2$, $var(K_j) = (j-1)^2 (1 - 1/j^2) / j^2 \to 1$ and $E\tilde{S}_n \sim \log n$. Thus the SLLN implies $\frac{\tilde{S}_n}{n} \overset{a.s.}{\to} 0$. But $\sum_j P(K_j = j) = \sum_j 1/j^2 < \infty$, so $P(K_j = j \text{ i.o.}) = 0$. That is, with probability 1, each realization of $K_1, K_2, \cdots$ ends with an infinite sequence of 1's. So $X_n \overset{a.s.}{\to} X$, where $P(1 \le X < \infty) = 1$, i.e. there exists a proper limiting stationary distribution; (4) fails.

Example 2 Suppose $K_1, K_2, \cdots$ are independent with $P(K_j = 1) = 1 - 1/j$ and $P\left(K_j = [\sqrt{j}]\right) = 1/j$. Then $var(K_j) \sim 1$, $E\tilde{S}_n \sim 2\sqrt{n}$, $\frac{\tilde{S}_n}{n} \overset{a.s.}{\to} 0$ and

$$\sum_j P\left(K_j = [\sqrt{j}]\right) = \sum_j 1/j = \infty.$$

Thus $P\left(K_j = [\sqrt{j}] \text{ i.o.}\right) = 1$ and hence $P(X_n \to \infty) = 1$; (6) fails.

2.2 iid examples

Even if we make the assumptions that $K_1, K_2, \cdots$ are iid with mean 1 and that $P(K_1 = 1) < 1$, we do not know if (6) always holds (of course, it does when $P(K_1 \ge 0) = 1$). Part of the problem is that the result $P(X_n \to 0 | X_1 = j) = (P(X_n \to 0 | X_1 = 1))^j$ is not generally true outside the branching process context. To see this, consider for example the case where $K_1, K_2, \cdots$ are iid with $P(K_1 = 1) = 1/6$, $P(K_1 = 2) = 3/4$, $P(K_1 = -2) = 1/12$. Since $EK_1 = 3/2 > 1$, Theorem 1(b) implies $q = P(X_n \to 0) < 1$. When $K_1 = 2$, let $\{X_{ni}, n \ge 2\}$ be the line of descent from "particle" i, $i = 1, 2$. Then

$$P\left(X_n \to 0 | X_1 = 2\right) = P\left(X_{n1} \to 0, \ X_{n2} \to 0\right)$$
$$+ 2P\left(X_{n1} \to \infty, \ \cup_{n \geq 2}\{X_{n1} + X_{n2} \leq 0\}\right)$$
$$= q^2 + 2\theta$$

and $\theta \geq P\left(X_{n1} \to \infty, X_{21} = 1, X_{22} = -2\right) = (1 - q)/72 > 0$. Hence

$$P\left(\text{extinction} | X_1 = 2\right) \neq P(\text{extinction} | X_1 = 1)^2.$$

Condition (8) is trivially satisfied with $C = 0$ in the branching process case. It can also be shown to hold for instance in the case where $K_1 + 1$ is Poisson with parameter 2, so that $\sum_{i=1}^{j} K_i + j$ is Poisson with parameter $2j$:

Lemma 1 *If Y is Poisson with parameter $2j$ then*

$$E(j - Y | Y - j \leq 0) \leq 2 \quad \text{for } j = 1, 2, \cdots. \tag{9}$$

Proof Write $\alpha_i = P(Y = i)$ and $A_i = P(Y \leq i)$. Then

$$E(Y I(Y \leq j)) = \sum_{i=0}^{j} i \alpha_i = 2j A_{j-1},$$

so

$$E(j - Y | j - Y \geq 0) = j - 2j \frac{A_{j-1}}{A_j}$$
$$= j\left(\frac{2\alpha_j}{A_j} - 1\right) \tag{10}$$

after a little algebra. Now

$$A_j = \alpha_j\left(1 + \frac{1}{2} + \frac{1}{4}\frac{j-1}{j} + \frac{1}{8}\frac{(j-1)(j-2)}{j^2} + \cdots + \frac{(j-1)!}{2^j j^{j-1}}\right)$$
$$= \alpha_j\left(1 + \frac{1}{2} + \frac{1}{4}\left(1 - \frac{1}{j}\right) + \frac{1}{8}\left(1 - \frac{1}{j}\right)\left(1 - \frac{2}{j}\right) + \cdots\right).$$

Using the fact that

$$\left(1 - \frac{1}{j}\right)\left(1 - \frac{2}{j}\right)\cdots\left(1 - \frac{k}{j}\right) \geq 1 - \frac{k(k+1)}{2j}, \quad k = 1, 2, \cdots, j_0,$$

with $j_0 = [\sqrt{2j} - 1]$, we get

$$A_j \geq \alpha_j\left(1 + \frac{1}{2} + \frac{1}{4} + \cdots + \frac{1}{2^{j_0+1}} - \frac{1}{4j}\left(1 + \frac{3}{2} + \frac{6}{4} + \cdots + \frac{j_0(j_0+1)}{2^{j_0}}\right)\right)$$
$$\geq \left(\frac{1 - 1/2^{j_0+2}}{1 - 1/2} - \frac{2}{j}\right)$$

using $\frac{2x}{(1-x)^3} = \sum_{m \geq 1} m(m+1)x^m$. Thus

$$A_j \geq 2\alpha_j\left(1 - \frac{1}{2^{j_0+2}} - \frac{1}{j}\right)$$

and so from (10)

$$E(j - Y | j - Y \geq 0) \leq j \left(\frac{1}{1 - 1/2^{j_0+2} - 1/j} - 1 \right)$$
$$\leq 1.5j \left(\frac{1}{2^{j_0+2}} + \frac{1}{j} \right)$$

for $j \geq 5$. It follows that Lemma 1 holds for $j \geq 13$. It is easy to verify the Lemma directly using (10) for $j \leq 12$.

Condition (8) is also satisfied in the case where $K_1 + rp - 1$ is Binomial with parameters r and p (we write $Y \sim \mathcal{B}(r, p)$) with rp integral, so that $\sum_{i=1}^{j} K_i + j(rp - 1) \sim \mathcal{B}(jr, p)$:

Lemma 2 *If Y is Binomial with parameters jr and p then*

$$E(j(rp - 1) - Y | Y - j(rp - 1) \leq 0) = O(1) \quad for \ j = 1, 2, \cdots . \tag{11}$$

Proof It can be shown as in the proof of Lemma 1 that

$$E(Y - r_j | Y - r_j \leq 0) = \frac{r_j P(Y \leq r_j) - E(Y I(Y \leq r_j))}{P(Y \leq r_j)}, \tag{12}$$

where $r_j = j(rp - 1)$ and $Y \sim \mathcal{B}(jr, p)$. Some algebra shows that the numerator of (12) is

$$A_j = r_j P(Y \leq r_j) - jrp P(Y_1 \leq jr - 1)$$
$$= p(jrq + j) P(Y_1 = r_j - 1) - j P(Y_1 \leq r_j - 1),$$

where $Y_1 \sim \mathcal{B}(jr - 1, p)$ and $q = 1 - p$. Now using the approximation (9.7.11) from Hald (1981) we get $A_j = P(Y_1 = r_j - 1) \cdot j A_j'$, where

$$A_j' = \frac{1}{j} a_j + \frac{1}{j} b_j + \frac{1}{j} O(j^{-1}), \tag{13}$$

while

$$a_j = \frac{j}{j - p + 1} (rq + 1) pq \to (rq + 1) pq$$

and

$$b_j = \frac{j}{j - p + 1} \cdot \frac{jp(rq + 1)}{j - p + 1} \cdot \frac{qj(rp - 1) - q}{j - p + 1} \to pq(rq + 1)(rp - 1).$$

It follows that

$$j A_j' \to rp^2 q(rq + 1). \tag{14}$$

It also follows from (9.7.11) of Hald (1981) that

$$P(Y \leq r_j) = P(Y_1 = r_j - 1) A_j'',$$

where $\lim_{j \to \infty} A_j'' = \frac{p^2 q}{rp - 1} + p(rq + 1) > 0$, which together with (12)-(14) proves Lemma 2.

2.3 More general results

The last two examples suggest that a large deviation result might be valid. This turns out to be correct: (8) is implied by

$$\phi(t) = E e^{-K_1 t} < \infty \text{ for all } t > 0. \tag{15}$$

This result is proved in Quine and Szczotka (1994).

Finally we note that by adapting its proof, it is possible to reduce the conditions of Theorem 2 to the following, involving the assumption

$$\mathbf{A} : E\left(X_{n+1}|\mathcal{F}_n\right) = a X_n^+, \ n = 0, 1, 2, \cdots,$$

where $\mathcal{F}_n$ is the σ-field generated by $X_1, \cdots, X_n$ and $X_n^+ = \max(X_n, 0)$ ($\mathbf{A}$ is satisfied for example if the K_i's are independent with mean a).

Theorem 3 *Sufficient conditions for (6) to hold are:*

(a) A holds,

(b) $\sup_{r \geq 0} -E\left(K_{r+1} + \cdots + K_{r+j}|K_{r+1} + \cdots + K_{r+j} \leq 0\right) \leq C$ *for all j and*

(c) there exist positive numbers $\alpha_1, \alpha_2 \cdots$ with $\sum \alpha_i < \infty$ such that

$$\sup_{r \geq 0} P(K_{r+1} + \cdots + K_{r+i} > 0) \leq 1 - \alpha_i, \ i = 1, 2, \cdots.$$

References

FELLER, W. *An introduction to probability theory and its applications* vol. 2. Wiley, New York, 1971.

HALD, A. *Statistical theory of sampling inspection by attributes.* Academic Press, London, 1981.

QUINE, M. P. AND SZCZOTKA, W. Generalisations of the Bienaymé-Galton-Watson branching process via its representation as an imbedded random walk. To appear in *Ann. Appl. Probab.* **4**, 1994.

SANKARANARAYANAN, G. *Branching Processes and its Estimation Theory.* Wiley, New York, 1989.

CRITICAL GENERAL BRANCHING PROCESSES WITH LONG-LIVING PARTICLES *†

TOPCHIJ V.A.,

Inst. of Information Technology and Applied Mathem.
of Siberian Branch of Russian Academy of Sciences,
Omsk, Russia.

Abstract

Various limit theorems for the critical Crump–Mode–Jagers process $Z(t)$ with long-living particles (non-extinction probabilities $P\{Z(t) > 0\}$ and $P\{X(t) > 0\}$ are equivalent, as $t \to \infty$, where the process $X(t)$, $t \geq 0$ are defined, as a number of particles in the entire process $\{Z(t), t \geq 0\}$ with the life-length of more then or equal to t) are discussed. These include asymptotics of non-extinction probabilities, convergence of deviations probabilities for $Z(t)$ (existence of a limit $\lim_{t\to\infty} \psi_1(t)\mathbf{P}\{Z(t) > x\psi_3(t)\}$ for some functions $\psi_1(t)$ and $\psi_3(t)$) and joint deviations probabilities for $Z(t)$ and $Y(t)$ - the total number of particles that have been born up to t. It is a summary of some previous works difficult of access.

CRITICAL GENERAL BRANCHING PROCESS, NON-EXTINCTION PROBA-BILITY, LIMIT THEOREMS FOR DEVIATIONS

AMS 1991 SUBJECT CLASSIFICATION: 60J80

1 Introduction

The general branching process (Crump–Mode–Jagers process, see [1]) $Z(t)$, $t \geq 0$, may be described as the evolution of some population of particles (t denoting time). The evolution of each individual is independent and identically distributed. They are determined by the distribution of stochastic process $\{\eta, N(t); \ t \geq 0\}$, where η is interpreted as a life-length, and a point process $N(t)$ is a general number of offsprings generated by a particle over time t from its birth.

Let $N = \lim_{t\to\infty} N(t) = N(\eta)$; $a(i)$, $i = 1,\ldots,N$ be the ages of particles at the moments of birth of the i-th offspring, $Z(t)$ be the number of particles existing but not dying at the instant t and $Y(t)$ be the total number that have been born up to t. We investigate the critical case, i.e., $\mathbf{E}N = 1$.

---

*Supported by Russian Fund of Fundamental Investigations 93-01-01471.

†Postal adress: Ul.Lermontova, 130, kv. 68, Omsk 644001, RUSSIA

The author wishes to thank Professor C.C.Heyde for the great edit help

All properties of Markov processes $Z(t)$ (N and η are independent, $N(t) = 0$, for $\eta < t$, and $\eta = 1$, or η is exponentially distributed) are described in terms of a generating function for N and its iterations. Properties of Bellman–Harris processes $Z(t)$ (N and η are independent, $N(t) = 0$, for $\eta < t$, and η is arbitrary distributed) are essentially dependent on generating function for N and $q(t) = \mathbf{P}\{\eta > t\}$. Properties of general processes $Z(t)$ are essentially dependent on $A(t)$ in addition to the last two parametric families, where $A(t) = max\{1, \mathbf{E}\{\sum_{i=1}^{N} a(i); \eta \leq t \}\}$, if $A(t) \rightarrow \infty$ for $t \rightarrow \infty$, or $A(t) = \mathbf{E}\sum_{i=1}^{N} a(i) = a$ otherwise. For Bellman–Harris processes $A(t)$ is a function (integral) of $q(t)$, but for general processes they are almost independent families and these effects are very interesting.

One of the main ideas of all the author's investigations is that of constructing suitable families of branching processes on a common probability space.

Let us define the process $X(t)$, $t \geq 0$, as the number of particles in the entire process $\{Z(t), t \geq 0\}$ with life-length more then or equal to t. Then (using a common probability space) the process $Z(t)$ may be called with long-living particles by chance probabilities $\mathbf{P}\{Z(t) > 0\}$ and $\mathbf{P}\{X(t) > 0\}$ are one and the same order (may be equivalent) for $t \rightarrow \infty$. Last effects are impossible in Markov case. Many asymptotic properties of $X(t)$ and the process $Z(t)$ with long-living particles coincide.

2 Non-extinction probability. Definition of process with long-living particles

Let $\mathcal{L}(\beta, d)$ be a family of regularly varying functions at point d ($d = 0$, or $d = \infty$), i.e. if $g(x) \in \mathcal{L}(\beta, d)$, then $g(x) = x^{\beta} l(x)$, where $l(x)$ is a slowly varying function at point d. Let $\mathcal{L}_c$ be a family of asymptotically constant functions at infinity, i.e. if $g(x) \in \mathcal{L}_c$; then there exists finite $\lim_{x \to \infty} g(x)$. Let $\mathcal{L}$ be a family of uniformly smooth at infinity functions, i.e. if $g(x) \in \mathcal{L}$, then $\lim_{t \to \infty, \delta \to 1} g(\delta x) g^{-1}(x) = 1$. Also, put $Q(t) = \mathbf{P}\{Z(t) > 0\}$; $F(t) = \mathbf{P}\{\eta \leq t\}$; $f(x) = \mathbf{E}x^{N}$; $q(t, y) = \mathbf{E}\{(1 - y)^{N}; \eta > t\}$; $\mu = \mathbf{E}\eta$; $G(y) = f(1 - y) - 1 + y$; $E(t), M(t)$ and $L(t)$ are the solutions of equations

$$\int_{E(t)}^{1} G^{-1}(x)dx = \int_{0}^{t} A^{-1}(s)ds - \int_{0}^{t} q(s, E(s))A^{-1}(s)G^{-1}(E(s))ds, \qquad (1)$$

$$\int_{0}^{t} A^{-1}(s)ds = \int_{M(t)}^{1} G^{-1}(x)dx \qquad (2)$$

and $q(t, L(t)) = G(L(t));_1 q(t, y) = q(t, y) + G(M(t))$.

Asymptotics of the non-extinction probability for critical processes were described by many authors (Kolmogorov [2] – first result for Galton–Watson processes; Zolotaryev [3], Slack [4] – infinite second moment for N; Goldstein [5] – new method of investigation; Vatutin [6] – generalization for Bellman–Harris processes and others) step by step together with generalization of branching schemes.

Most of the known results about the non-extinction probability for general or the simplest critical processes may be obtained from the following theorem:

Theorem 1 [7,8].Let $Z(t)$ be a general critical branching process,

$$_1 q(t, g(t)L(t)) \in \mathcal{L}$$

for every $g(t) \in \mathcal{L}_c, A(t) \in \mathcal{L}$ and suppose that the conditions

$$\lim_{c,y \searrow 0} G(cy)(cG(y))^{-1} = 0,$$

$$\lim_{\varepsilon \searrow 0, t \to \infty} \int_0^{\varepsilon t} A^{-1}(s)ds \left(\int_0^t A^{-1}(s)ds \right)^{-1} = 0$$

and for every $\varepsilon \in (0,1)$ as $t \to \infty$

$$\mathbf{E}\{N - N(\varepsilon t); \eta \le t\} = o(G(E(t)))E^{-1}(t)$$

are true.

Then for $t \to \infty$

$$Q(t) \sim E(t).$$

The last term in the right side of Definition (1) may be infinitesimal with respect to the left side, then the function $E(t)$ in the assertion of Theorem 1 may be replaced to $M(t)$ (Definition (2)) . In reverse situation $E(t)$ may be replaced by $L(t)$. In the first case the behaviour of the non-extinction probability is the same as for a Markov process and may be described in terms of iterations of the generating function $f(x)$. In the second case we call $Z(t)$ the process with long-living particles.

Direct analysis of a wider class of process leads to the following theorem for the process with long-living particles without using Markov component $E(t)$.

Theorem 2 [9]. Let $Z(t)$ be a general critical branching process,

$$q(t, g(t)L(t)) \in \mathcal{L}$$

for every $g(t) \in \mathcal{L}_c$ and for $t \to \infty$ the condition

$$A(t)L(t)G^{-1}(L(t)) = o(t), \tag{3}$$

is true;

then

$$Q(t) \sim L(t).$$

The common probability space method allows one to transfer the birth of offspring to age 0, i.e. all offspring are born at 0, the initial time moment. After that all conditions of Theorem 2 are true for a new process and $L(t)$ is the same in both cases and it was used in the introduction. Condition (3) extends the condition $E(t) = o(L(t))$.

For Bellman–Harris processes N and η are independent and $q(t, y) \sim q(t)$, as $t \to \infty$, and $L(t)$ may be defined by the equation $G(L(t)) = q(t)$, or equivalently $L(t) = G_{-1}(q(t))$. Then $A(t) \sim \int_0^t t dF(t)$ and it follows from Theorem 2 that:

Corollary 1. Let $Z(t)$ be a critical Bellman–Harris process, $q(t) \in \mathcal{L}$ and the condition

$$\int_0^t u dq(u) G_{-1}(q(t)) q^{-1}(t) = o(t),$$

holds, as $t \to \infty$.

Then for $t \to \infty$

$$Q(t) \sim G_{-1}(q(t)).$$

3 Generating functions for process with long-living particles. Limit theorems

The main ideas of investigations of non-extinction probabilities allow one to obtain the asymptotic behaviour of $Q(t,z) = 1 - \mathbf{E}z^{Z(t)}$, as $t \to \infty$ and z is fixed or $z = z(t)$ and tends to 1. This leads to wide classes of limit theorems in different forms. We discuss here only the process with long-living particles and give asymptotics of generating functions and mainly deviations probabilities. For generating functions the particular form of $G(y)$ does not matter, but for probabilities one can show that non-trivial normalized limits are existent if and only if $G(x) \in \mathcal{L}(1 + \alpha, 0)$ for some $\alpha \in (0, 1]$. So for compactness we write the last condition for $G(x)$ and analogous conditions for functions in corresponding theorems and mark them $(^*)$.

Let $L(t,z)$ be the solution of equation $(1 - z)q(t, L(t,z)) = G(L(t,z))$.

Theorem 3 [9]. *Let $Z(t)$ be a general critical branching process and the condition of Theorem 2,*

$$\lim_{t,w \to \infty} \mathbf{E}\{1 - \mathbf{P}^N\{Z(t) = 0\}; \eta > wt\}q^{-1}(t) = 0,$$

$$\lim_{\delta \nearrow 1, t \to \infty} \int_{\delta t}^{t} Q(t - u)d\mathbf{E}N(u)q^{-1}(t) = 0,$$

hold;
  then for $t \to \infty$

$$Q(t,z) \sim L(t,z). \tag{4}$$

Here, as in Theorem 2, the common probability space method allows us to transfer the birth of offsprings at the age 0, i.e. all offsprings are born at 0, the initial time moment. After that, for the new process all conditions of Theorem 3 hold and $L(t,z)$ is the same in both cases, but the second one coincides with $X(t)$ from the introduction, i.e. the asymptotics of $L(t,z)$ dependents on the life-lengths of particles in the entire process $\{Z(t), t \geq 0\}$.

From the last theorem it follows that:

Theorem 4 $(^*)$ [9]. *Let $Z(t)$ be a general critical branching process, satisfying the condition of Theorem 3, $G(x) \in \mathcal{L}(1 + \alpha, 0)$ for some $\alpha \in (0, 1]$ and the condition $q(t, cL(t)) \sim \psi_0(c)q(t)$ holds for some function $\psi_0(c)$ for every $c \in (0, 1]$ as $t \to \infty$.*
  Then for any natural k the limits

$$\lim_{t \to \infty} \mathbf{P}\{Z(t) = k \mid Z(t) > 0\}$$

exist and their generating function $1 - \psi(z)$ is the solution of equation

$$\psi^{1+\alpha}(z) = (1 - z)\psi_0(\psi(z)).$$

Note that for Bellman–Harris processes ($\psi_0(c) \equiv 1$ and $\psi(z) = (1 - z)^{1/(1+\alpha)}$) the result here was first published by Vatutin [6].

Now we describe the conditions for existence of a non-trivial limit

$$u(x) = \lim_{t \to \infty} \psi_1(t)\mathbf{P}\{Z(t) > x\psi_3(t)\}$$

for some functions $\psi_1(t)$ and $\psi_3(t)$. For that purpose formula (4) for $z = z(t)$ in several regions (corresponding to different $\psi_i(t)$) is obtained.

Theorem 5 [9]. Let $Z(t)$ be a general critical branching process, $q(t) \in \mathcal{L}, z(t) = 1 - yu(t)$ for some $u(t) \in \mathcal{L}$ such that for $t \to \infty$ $u(t) \searrow 0$ and

$$A(t)L(t, z(t))((1 - z(t))q(t))^{-1} = o(t). \tag{5}$$

Suppose also that

$$\lim_{w,t\to\infty} L(t, z(t))\mathbf{E}\{N; \eta > wt\}G^{-1}(L(t, z(t))) = 0 \tag{6}$$

and if $\mathbf{E}\eta = \infty, a < \infty$, then

$$\lim_{t\to\infty,\delta\to1-0} \int_0^{(1-\delta)t} q(u)du\mathbf{E}(N(t) - N(\delta t))q^{-1}(t) = 0 \tag{7}$$

or otherwise

$$\lim_{t\to\infty,\delta\to1-0} \sup_{0<\varepsilon<1-\delta} (1 + \frac{q(\varepsilon t)}{1 - \mathbf{E}N(\varepsilon t)}) \frac{\mathbf{E}(N(t) - N(\delta t))}{q(t)} = 0. \tag{8}$$

Then for $t \to \infty$,

$$Q(t, z(t)) \sim L(t, z(t)) = G_{-1}((1 - z(t))q(t)). \tag{9}$$

For Bellman–Harris processes the following result may be obtained from Theorem 5.

Corollary 2 [9]. Let $Z(t)$ be a critical Bellman–Harris process, $q(t) \in \mathcal{L}$ $z(t) = 1-yu(t)$ for some $u(t) \in \mathcal{L}$ such that for $t \to \infty$ $u(t) \searrow 0$ and the condition (5),

$$\lim_{w,t\to\infty} q(tw)/q(t) = 0,$$

as $\mathbf{E}\eta = \infty$, holds.

Then (9) is true.

From Theorem 5 the following result can be deduced.

Theorem 6 (*) [9].Let $Z(t)$ be a general critical branching process, $G(x) \in \mathcal{L}(1 + \alpha, 0)$ for some $\alpha \in (0, 1]$, $q(t) \in \mathcal{L}, z(t) = exp\{\lambda\psi_3^{-1}(t)\}$ for some $\psi_3(t)$ such that $\sup_{x>c} \sup_{2x>t>x/2} \psi_3(t)/\psi_3(x) < \infty$, is true for some $c > 0$ for $t \to \infty$ $\psi_3(t) \to \infty$ and (5), (6) and either (7) or (8) is true.

Then for $x(t) \to \infty, x(t) = O(\psi_3(t))$, as $t \to \infty$, it follows that

$$\mathbf{P}\{Z(t) > x(t)\} \sim \Gamma^{-1}(\alpha/(1 + \alpha))G_{-1}(x^{-1}(t)q(t)). \tag{10}$$

Note that for Bellman–Harris processes a result generalizing one of Vatutin [10] follows from Corollary 2.

Theorem 5 corresponds to the case $\psi_3(t) = o(G^{-1}(M(t))q(t))$. In another case, when $\psi_3(t)$ is the same order as $G^{-1}(M(t))q(t)$, holds.

Theorem 7 [9]. Let $Z(t)$ be a general critical branching process, $G(x) \in \mathcal{L}(1 + \alpha, 0)$, $q(t) \in \mathcal{L}(\beta, \infty)$ for some $\alpha \in (0, 1]$, $\beta \in (0, (1 + \alpha)/\alpha]$, $A(t)$ be a slowly varying function at infinity, for $t \to \infty$ $u(t) \sim G(M(t))q^{-1}(t), u(t) \searrow 0, M(t) = o(L(t))$, for any $\varepsilon \in (0, 1)$

$$\mathbf{E}\{N(t) - N(\varepsilon t); \eta \leq t\} = o(A(t)/t),$$

$$\lim_{w,t\to\infty} tA^{-1}(t)\mathbf{E}\{N;\eta > wt\} = 0$$

and eihter (7) or (8) holds.
  Then the limit

$$\lim_{t\to\infty} \mathbf{P}\{u(t)Z(t) > x\}M^{-1}(t) = u(x) \tag{11}$$

exists and has the Laplace transform $_0U(\lambda) = U(\lambda)\lambda^{-1}$, where $U(\lambda)$ is a solution of the differential equation

$$U(\lambda) + \lambda U'(\lambda)(\alpha\beta - 1 - \alpha) = U^{1+\alpha}(\lambda) - \lambda \tag{12}$$

with initial conditions $U(0) = 1$ for $\beta \geq 1$ and $U(0) = 0, U'(0) = \alpha^{-1}(1 - \beta)^{-1}$ for $\beta < 1$.
  For Bellman–Harris processes last result is similar to Vatutin [11]. But generalizing the scheme widens the class of limiting functions. In particular, for Bellman–Harris processes the cases $\beta < 1$ and $A(t)$ is a slowly varying function at infinity, or $\mathbf{E}\eta = \infty, a < \infty$ are impossible.

4 Limit theorems for common deviations for $Z(t)$ and $Y(t)$

Here we describe conditions for existence of the non-trivial limit

$$u(x,y) = \lim_{t\to\infty} \psi_1(t)\mathbf{P}\{Y(t) > x\psi_2(t), Z(t) > y\psi_3(t)\}$$

for some functions $\psi_1(t), \psi_2(t)$ and $\psi_3(t)$. In the previous section the case $\psi_3(t) \equiv 0$ (see (10—11)) was investigated. For completeness results for deviations of Y(t) on the set $\{Z(t) = 0\}$ are given .
  Theorem 8 (*) [12]. Let $Z(t)$ be a general critical branching process, $G(x) \in \mathcal{L}(1+\alpha, 0)$ for some $\alpha \in (0,1]$, $\mathbf{P}\{\eta = 0\} = 0, Q(t) \in \mathcal{L}, x(t) = o(G^{-1}(Q(t)))$, as $x(t) \to \infty$ for $x \to \infty$.
  Then

$$\mathbf{P}\{Y(t) > x(t); Z(t) = 0\} \sim \Gamma^{-1}(\alpha/(1+\alpha))G_{-1}(x^{-1}(t)).$$

Here the long-living particles condition does not matter and the assertion has no analogy.
  Theorem 8 corresponds to the case $x(t) = o(G^{-1}(Q(t)))$. If $x(t)$ is the same order as $G^{-1}(Q(t))$,the long-living particles condition is important and the following result holds.
  Theorem 9 (*) [13,12]. In the conditions of Theorem 2, as $G(x) \in \mathcal{L}(1 + \alpha, 0)$ for some $\alpha \in (0,1]$, and there exists a function $\psi_0(c)$ such that for any $c \in (0,1]$ as $t \to \infty$ the condition $q(t, cL(t)) \sim \psi_0(c)q(t)$ holds,
  the limit

$$\lim_{t\to\infty} \mathbf{P}\{q(t, L(t))Y(t) > x; Z(t) = 0\}L^{-1}(t) = v(x)$$

exists and has the Laplace transform $V(\lambda)$ which is the unique solution of the equation

$$\lambda = (\lambda V(\lambda) + 1)^{\alpha+1} - \psi_0(\lambda V(\lambda) + 1).$$

Note that $v(x)$ may be found only for simple $\psi_0(c)$. Let $\psi_0(c) \equiv 1$, then $Z(t)$ may be Bellman–Harris process and the following result holds.

Corollary 3 [13,12]. In the conditions of Theorem 2, as $\psi_0(c) \equiv 1$ holds, then

$$v(x) = \Gamma^{-1}(\alpha/(1+\alpha))(e^{-x}x^{-(1+\alpha)^{-1}} - \Gamma(\alpha/(1+\alpha), x))$$

holds, where

$$\Gamma(\gamma, x) = \int_x^\infty e^{-t}t^{\gamma-1}dt$$

is the incomplete gammafunction and $\Gamma(\gamma) = \Gamma(\gamma, 0)$ is gammafunction.

For Bellman–Harris processes corollary 3 is more general than the result which may be obtained from Vatutin [14].

For investigating $\mathbf{P}\{Y(t) > \psi_2(t)x, Z(t) = k\}$ define

$$V_t(\lambda, z) = L^{-1}(t)\lambda^{-1}\mathbf{E}\{z^{Z(t)}(1 - \exp\{-\lambda q(t, L(t))Y(t)\}; Z(t) > 0\}.$$

Theorem 10 (*) [15]. In the conditions of Theorem 2, as $F(0) = 0$ the limit

$$V(\lambda, z) = \lim_{t\to\infty} V_t(\lambda, z)$$

exists and equals

$$V(\lambda, z) = \lambda^{-1}({}_0V(\lambda, z) - \lambda V(\lambda) - \psi(z)),$$

where $V(\lambda)$ from (12), $\psi(z)$ from Theorem 4 and ${}_0V(\lambda, z)$ is the unique solution of the equation

$${}_0V^{\alpha+1}(\lambda, z) - \lambda - \psi_0({}_0V(\lambda, z))(1 - z) = 0.$$

As $\psi_0(c) \equiv 1$ and $\gamma = (\alpha + 1)^{-1}$ then

$$V(\lambda, z) = \lambda^{-1}(1 + (1 - z + \lambda)^\gamma - (1 - z)^\gamma - (1 + \lambda)^\gamma).$$

The last assertion of Theorem 10 is connected with Vatutin's results [14] for Bellman–Harris processes and involves

$$\mathbf{P}\{Z(t) = k \mid Z(t) > 0\} \sim (-1)^k\gamma(\gamma - 1)\ldots(\gamma - k + 1)/k!,$$

$$\mathbf{P}\{Y(t)q(t) > x \mid Z(t) = k\} \sim \Gamma^{-1}(k - \gamma)\int_x^\infty e^{-u}u^{k-\gamma-1}du,$$

where $\Gamma(\gamma)$ is the gammafunction and k is a natural number.

The next assertion has no analogue.

Theorem 11 (*) [15]. Suppose that the conditions of Theorem 5 with $z(t) = exp\{\lambda_2\psi_3^{-1}(t)\}$ for some $\psi_3(t) \in \mathcal{L}$, $\psi_3(t) \nearrow \infty$ holds, $G(x) \in \mathcal{L}(1 + \alpha, 0)$ for some $\alpha \in (0, 1]$, $\psi_2(t) = \psi_3(t)q^{-1}(t)$, and $\psi_1(t) = G_{-1}^{-1}(\psi_2^{-1}(t))$.

Then the limit

$$u(x, y) = \Gamma^{-1}(\alpha\gamma)\min\{x^{-\gamma}, y^{-\gamma}\}$$

exists.

Theorem 11 may be written in the following form:

Corollary 4 [15]. In the conditions of Theorem 11

$$\mathbf{P}\{Y(t) > x(t), Z(t) > y(t)\} \sim \Gamma^{-1}(\alpha\gamma)G_{-1}(\min\{x^{-1}(t), y^{-1}(t)q(t)\})$$

hods, where $x(t) \to \infty, y(t) \to \infty, x(t) = O(\psi_2(t)), y(t) = O(\psi_3(t))$, as $t \to \infty$.

References

[1] JAGERS, P. (1975) *Branching processes with biological applications.* Wiley, New York.

[2] KOLMOGOROV, A. (1938) Zur Lösung einer biologischen Aufgabe. *Izvestiya Nauchno-issledov. Inst. Matematiki i Mechaniki pri TGUniv.* 2, 1–6.

[3] ZOLOTARJOV,V.M. (1957) More precise theorems for branching random processes. *Teoriya Veroyatnostei i ee Primeneniya.* 2, 256–266.

[4] SLACK,R.S. (1968) A branching processes with mean one and possibly infinite variance. *Z.Wahrscheinlichkeitstheorie verw.* Geb. 9, 139–145.

[5] GOLDSTEIN, M. (1971) Critical age-dependent branching processes: single and multitype. *Z.Wahrscheinlichkeitstheorie verw.* Geb. 17, 74–88.

[6] VATUTIN,V.A. (1979) New limit theorem for critical Bellman–Harris branching process. *Matematicheskii Sbornik.* 109, 440–452.

[7] TOPCHIJ,V.A. (1987) Properties of non-extinction probability for general branching processes under the weak conditions. *Sibirskii Matematicheskii Zhurnal.* 28, 178–192.

[8] TOPCHIJ,V.A. (1987) Generalization of results about non-extinction probability for general branching processes. *Stochastic Models and Information Systems.* Novosibirsk, Computer Center, 143–179.

[9] TOPCHIJ,V.A. (1988) Limit theorems for a critical general branching processes with long-living particles. *Stochastic and Deterministic Models of Complicated Systems.* Novosibirsk, Computer Center, 114–153.

[10] VATUTIN,V.A. (1986) Critical Bellman-Harris branching process,beginning from large number of particles. *Matematicheskii Zametki.* 40, 527–541.

[11] VATUTIN,V.A. (1987) Asymptotical properties of critical Bellman-Harris branching process,beginning from large number of particles.*Problems of Stochastic Models Stability.* Moscow, VNIISI,8–15.

[12] TOPCHIJ,V.A. (1988) Temperate deviations for total number of particles in a critical branching processes. *Teoriya Veroyatnostei i ee Primeneniya.* 33, 406–409.

[13] TOPCHIJ,V.A. (1990) Asymptotic of probability deviations for total number of particles in a critical branching processes. *Stochastic Models of Complicated Systems.* Novosibirsk, Computer Center. 88–94.

[14] VATUTIN,V.A. (1986) Critical Bellman–Harris branching process, with final class of particles. *Teoriya Veroyatnostei i ee Primeneniya.* 31, 491–452.

[15] TOPCHIJ,V.A. (1991) Joint distributions for critical general branching processes with one type of long-living particles. *Sibirskii Matematicheskii Zhurnal.* 32, 153–164.

CRITICAL BRANCHING PROCESSES WITH RANDOM MIGRATION

GEORGE P. YANEV *AND NICKOLAY M. YANEV*
Bulgarian Academy of Sciences, Sofia

Abstract

A new class of branching processes allowing a random migration component in every generation is considered: with probability p two types of emigration are possible - a random number of families and a random number of individuals, or with probability q there is not any migration (i.e. the process develops like a Bienaymé-Galton-Watson process), or with probability r a state-dependent immigration of new individuals is available, $p + q + r = 1$. The coresponding processes stopped at zero are studied in the critical case and the asymptotic behaviour of the non-extinction probability is obtained (depending on the range of an extra critical parameter).
branching process;   random migration;   extinction
60J80

1   Introduction

The main purpose of this paper is to introduce a new class of branching processes in which evolution is not isolated and which admits a random migration component. The development of the process depends on relative sizes of migration and branching parameters and the results obtained are quite different from those in the classical Bienaymé-Galton-Watson process (BGWP).

The processes considered can be defined as follows. Let us have on some probability space three independent sets of integer-valued random variables, i.i.d. in each set, $X = \{X_{i,t}\}$, $\eta = \{(\eta_{1t}, \eta_{2t})\}$, $I = \{(I_t, I_t^0)\}$. Then define

$$Y_t = \left(\sum_{i=1}^{Y_{t-1}} X_{i,t} + M_t \right)^+ \qquad t = 1, 2, ..., \quad Y_0 \geq 0 , \qquad (1.1)$$

where

$$M_t = \begin{cases} -\left(\sum_{i=1}^{\eta_{1t}} X_{i,t} + \eta_{2t} \right) \mathbb{I}_{\{Y_{t-1} > 0\}} & \text{with probability } p, \\ 0 & \text{with probability } q, \\ I_t \mathbb{I}_{\{Y_{t-1} > 0\}} + I_t^0 \mathbb{I}_{\{Y_{t-1} = 0\}} & \text{with probability } r, \qquad p+q+r=1, \end{cases}$$

*Postal address for both authors: Department of Probability and Statistics, Institute of Mathematics, Bulgarian Academy of Sciences, 1113 Sofia, Bulgaria.
Supported by the National Foundation for Scientific Investigations, grant MM-4/91.

and Y_0 is independent of X, η, I. As usual $a^+ = \max(0, a)$.

The definition (1.1) admits the following interpretation. As usual, $X_{i,t}$ is the offspring in the t-th generation of the i-th individual which exists in the $(t-1)$-th generation. Then in the t-th generation the following three situations are possible: **(i)** with probability $p : \eta_{1t}$ families emigrate which give $\sum_{i=1}^{\eta_{1t}} X_{i,t}$ emigrants (*family emigration*) , in other words, η_{1t} individuals are eliminated in the $(t-1)$-th generation (before the reproduction) and do not take part in the further evolution, and additionally after the reproduction in the $(t-1)$-th generation η_{2t} individuals emigrate from the t-th generation who can be chosen randomly from different families (*individual emigration*); **(ii)** with probability q : the reproduction is according to the BGWP, i.e. without any migration; **(iii)** with probability r : a state-dependent *immigration* of new individuals is possible: I_t individuals *in the non-zero* states or I_t^0 *in the state zero.*

If $q = 1$ then from (1.1) it follows that $\{Y_t\}$ is a classical BGWP. If $r = 1$ we obtain the BGWP with two (possible dependent) immigration components depending on the states of the process. The case $r = 1$ and $I_t \equiv I_t^0$ a.s. is the well-known BGWP with immigration. The process (1.1) with $p = 1$ i.e. the process with emigration was studied for $\eta_{2t} \equiv 0$ a.s. by Vatutin (1977a) and Kaverin (1990) and for $\eta_{1t} \equiv 0$ a.s. by Grey(1988). Yanev and Mitov(1980, 1981, 1983) investigated (1.1) where a.s. $\eta_{1t} \equiv 1$, $\eta_{2t} \equiv 0$ and $I_t \equiv I_t^0$. Nagaev and Han(1980) and Han(1980) studied the case $0 \leq \eta_{1t} \leq N_1$, $\eta_{2t} \equiv 0$ and $I_t \equiv I_t^0 = \sum_{i=1}^{\eta_{3t}} X_{i,t}$ a.s. where η_{3t} is a r.v. Some results for the processes with $\eta_{1t} \equiv 0$ and $I_t \equiv I_t^0$ a.s. were announced by Yanev and Yanev(1991). Models with non-homogeneous migration, i.e. $p = p_t, q = q_t$ and $r = r_t$ were investigated by Yanev and Mitov(1985).

An important characteristic of the process $\{Y_t\}$ is the *life-period* $\tau = \tau(T)$ started at the moment $T \geq 0$, and defined by $Y_{T-1} = 0$, $Y_{T+n} > 0$, $0 \leq n < \tau$, $Y_{T+\tau} = 0$.

Further on we will also consider the process stopped at zero $\{Z_t\}$ defined by

$$Z_t \stackrel{d}{=} Y_{T+t} \mathbb{I}_{\{Z_{t-1} > 0\}}, \quad t = 1, 2, \ldots; \qquad Z_0 \stackrel{d}{=} Y_T > 0, \tag{1.2}$$

where T is the starting moment for a life-period i.e. $P(Y_T = n) = P(I_1^0 = n \mid I_1^0 > 0)$, $n \geq 1$.

Note that the state zero is an reflecting barrier for the process $\{Y_t\}$ and an absorbing barrier for the process $\{Z_t\}$. The distribution of the stay at zero is geometric with a parameter $a = P(Y_t = 0 \mid Y_{t-1} = 0) = 1 - rP(I_t^0 > 0) > 0$.

Processes stopped at zero were studied in the case $r = 1$ by Zubkov(1972), Vatutin (1977b), Seneta and Tavaré(1983) and Ivanoff and Seneta(1985). Yanev, Vatutin and Mitov(1986) investigated the process (1.2) with $\eta_{1t} \equiv 1$ and $\eta_{2t} \equiv 0$ a.s.

From (1.2) and the definition of τ it follows that

$$P(Z_t > 0) = P(\tau > t) = u_t , \qquad say. \tag{1.3}$$

The asymptotic behaviour of u_t was studied for the critical processes by Yanev and Mitov(1980,1983) and Yanev and Yanev(1993) in the case $\eta_{1t} \equiv 1$, $\eta_{2t} \equiv 0$. Now we will consider u_t in the most general situation (1.1). Note that the asymptotics of u_t is "the key" to obtaining limit theorems.

2 Equations and main results

In the sequel $F(s) = Es^{X_{i,t}}$ is the offspring p.g.f.; $H(s_1, s_2) = Es_1^{\eta_{1t}} s_2^{\eta_{2t}}$ is the p.g.f. of emigration; $G(s_1, s_2) = Es_1^{I_t} s_2^{I_t^0}$ is the p.g.f. of immigration, so $g(s) = G(s, 1)$ is the p.g.f. of immigration for positive states, $g_0(s) = G(1, s)$ is the p.g.f. of immigration in zero.

Assume that $h(s) = H(F^{-1}(s), s^{-1})$ is well defined for $0 < s \le 1$, $h(0) = 0$.
We will use the notation for $t = 1, 2, \ldots$

$$
\left.
\begin{aligned}
&\delta(s) = ph(s) + q + rg(s), \quad \Delta(s) = 1 - \delta(s) - r(1 - g_0(s)), \\
&F_t(s) = F(F_{t-1}(s)), \quad F_0(s) = s, \, F_t = F_t(0), \\
&\gamma_t(s) = \prod_{j=0}^{t-1} \delta(F_j(s)), \quad \gamma_t = \gamma_t(0), \, \gamma_0 = 1.
\end{aligned}
\right\}
\tag{2.1}
$$

It is convenient to introduce a random field $\mu_t(n) = \eta_{2t} - \sum_{i=1}^{n} X_{i,t}$, $t, n \ge 0$, and the functionals:

$$
\left.
\begin{aligned}
W(\zeta_{t-1}, s) &= E\{1 - F^{-(\eta_{1t} - \xi_{t-1})}(s) s^{-\eta_{2t}}\} \mathbb{I}_{\{\eta_{1t} \ge \xi_{t-1} > 0\}} + \\
&+ E\{1 - s^{-\mu_t(\xi_{t-1} - \eta_{1t})}\} \mathbb{I}_{\{\mu_t(\xi_{t-1} - \eta_{1t}) \ge 0, \xi_{t-1} > \eta_{1t}\}}, \\
W(\xi_{t-1}, 0) &= E\mathbb{I}_{\{\eta_{1t} \ge \xi_{t-1} > 0\}} + E\mathbb{I}_{\{\mu_t(\xi_{t-1} - \eta_{1t}) \ge 0, \xi_{t-1} > \eta_{1t}\}},
\end{aligned}
\right\}
\tag{2.2}
$$

where $s \ne 0$, $t = 1, 2, \ldots$ and ξ_{t-1} will be Y_{t-1} or Z_{t-1}.

Let $\Phi(t, s) = Es^{Y_t}$ and $\Psi(t, s) = Es^{Z_t}$ be the p.g.f. of the processes . Remember that in the case $q = 1$ they satisfy equations $\Phi(t, s) = \Phi(t - 1, F(s))$ and $\Psi(t, s) = \Psi(t - 1, F(s))$.

Theorem 2.1 *The p.g.f. $\Phi(t, s)$ and $\Psi(t, s)$ of the processes (1.1) and (1.2) satisfy the equations for $t = 1, 2, \ldots$ and $0 \le s \le 1$:*

$$
\begin{aligned}
\Phi(t, s) &= \delta(s)\Phi(t - 1, F(s)) + \Delta(s)\Phi(t - 1, 0) + pW(Y_{t-1}, s), &\tag{2.3} \\
\Psi(t, s) &= \delta(s)\Psi(t - 1, F(s)) + (1 - \delta(s))\Psi(t - 1, 0) + pW(Z_{t-1}, s). &\tag{2.4}
\end{aligned}
$$

Proof. From the definition (1.1) it follows that for $0 \le s \le 1$

$$
\begin{aligned}
\Phi(t, s) &= pEs^{\left(\sum_{i=1}^{(Y_{t-1} - \eta_{1t})^+} X_{i,t} - \eta_{2t}\right)^+} + qEs^{\sum_{i=1}^{Y_{t-1}} X_{i,t}} \\
&\quad + rEs^{\left(\sum_{i=1}^{Y_{t-1}} X_{i,t} + I_t \mathbb{I}_{\{Y_{t-1} > 0\}} + I_t^0 \mathbb{I}_{\{Y_{t-1} = 0\}}\right)}
\end{aligned}
\tag{2.5}
$$

$$
= p\Phi_{em}(t, s) + q\Phi_{nm}(t, s) + r\Phi_{im}(t, s), \quad \text{say.}
$$

Obviously, for $0 \le s \le 1$ one has

$$
\left.
\begin{aligned}
\Phi_{nm}(t, s) &= \Phi(t - 1, F(s)) \\
\Phi_{im}(t, s) &= \Phi(t - 1, F(s))g(s) + \Phi(t - 1, 0)(g_0(s) - g(s)).
\end{aligned}
\right\}
\tag{2.6}
$$

On the other hand, for $0 \leq s \leq 1$ one gets

$$
\begin{aligned}
\Phi_{\text{em}}(t,s) &= P(Y_{t-1} = 0) + P(\eta_{1t} \geq Y_{t-1} > 0) + \\
&+ P(\eta_{1t} < Y_{t-1}, \mu_t(Y_{t-1} - \eta_{1t}) \geq 0) + \\
&+ Es^{\sum_{i=1}^{Y_{t-1}-\eta_{1t}} X_{i,t} - \eta_{2t}} \mathbb{I}_{\{\eta_{1t} < Y_{t-1}, \mu_t(Y_{t-1} - \eta_{1t}) < 0\}}.
\end{aligned}
\tag{2.7}
$$

Replacing in a consecutive order

$$
\mathbb{I}_{\{\eta_{1t} < Y_{t-1}, \mu_t(Y_{t-1} - \eta_{1t}) < 0\}} = \mathbb{I}_{\{\eta_{1t} < Y_{t-1}\}} - \mathbb{I}_{\{\eta_{1t} < Y_{t-1}, \mu_t(Y_{t-1} - \eta_{1t}) \geq 0\}}
$$

$$
\mathbb{I}_{\{\eta_{1t} < Y_{t-1}\}} = 1 - \mathbb{I}_{\{\eta_{1t} \geq Y_{t-1} > 0\}} - \mathbb{I}_{\{\eta_{1t} \geq Y_{t-1} = 0\}}
$$

in (2.7) one obtains for $0 < s \leq 1$,

$$
\begin{aligned}
\Phi_{\text{em}}(t,s) &= \Phi(t-1, F(s))h(s) + \Phi(t-1,0)(1 - h(s)) + \\
&+ E\left(1 - F^{-(\eta_{1t}-Y_{t-1})}(s)s^{-\eta_{2t}}\right) \mathbb{I}_{\{\eta_{1t} \geq Y_{t-1} > 0\}} + \\
&+ E\left(1 - s^{-\mu_t(Y_{t-1} - \eta_{1t})}\right) \mathbb{I}_{\{\eta_{1t} < Y_{t-1}, \mu_t(Y_{t-1} - \eta_{1t}) \geq 0\}}.
\end{aligned}
\tag{2.8}
$$

From (2.7) and (2.2) we have

$$
\Phi_{\text{em}}(t,0) = \Phi(t-1,0) + W(Y_{t-1},0).
\tag{2.9}
$$

The relations (2.6) - (2.9) prove (2.3). Similarly one obtains (2.4).
Now iterating (2.3) and (2.4) it is not difficult to obtain that for $0 \leq s \leq 1$

$$
\begin{aligned}
\Phi(t,s) &= \Phi(0, F_t(s))\gamma_t(s) + \sum_{k=0}^{t-1} \Phi(t-1-k,0)\Delta(F_k(s))\gamma_k(s) + \\
&+ p\sum_{k=0}^{t-1} W(Y_{t-1-k}, F_k(s))\gamma_k(s)
\end{aligned}
\tag{2.10}
$$

and

$$
\begin{aligned}
\Psi(t,s) &= \Psi(0, F_t(s))\gamma_t(s) + \sum_{k=0}^{t-1} \Psi(t-1-k,0)(1 - \delta(F_k(s)))\gamma_k(s) + \\
&+ p\sum_{k=0}^{t-1} W(Z_{t-1-k}, F_k(s))\gamma_k(s).
\end{aligned}
\tag{2.11}
$$

From the definition (1.2) it follows that

$$
\Psi(0,s) = Es^{Z_0} = \frac{g_0(s) - g_0(0)}{1 - g_0(0)} = Q(s), \qquad \text{say.}
\tag{2.12}
$$

We will investigate the critical case under the following conditions:

$$
\left.
\begin{aligned}
&F'(1) = 1, \ 0 < F''(1) = 2b < \infty, \ \lambda = g'(1) < \infty, \ g_0'(1) < \infty, \\
&0 \leq \eta_{1t} \leq N_1, \quad 0 \leq \eta_{2t} \leq N_2, \quad N_1, N_2 < \infty.
\end{aligned}
\right\}
\tag{2.13}
$$

Let $E\eta_{1t} = e_1$, $E\eta_{2t} = e_2$. Then we define an extra critical parameter which play a important role in the behaviour of the processes by

$$\theta = \frac{r\lambda - p(e_1 + e_2)}{b} = \frac{E(M_t \mid Y_{t-1} > 0)}{\frac{1}{2}\mathrm{Var}X_{i,t}}.$$

Sometimes we will need the stronger conditions:

$$\left.\begin{array}{ll} 0 < \ \theta \leq 1 & EI_1 \log(1 + I_1) < \infty, \quad EX_{1,1}^2 \log(1 + X_{1,1}) < \infty, \\ \theta = 0 & EI_1 \log(1 + I_1) < \infty. \end{array}\right\} \tag{2.14}$$

The following result generalize that of Yanev and Mitov(1980, 1983) obtained in the case when a.s. $\eta_{1t} \equiv 1$, $\eta_{2t} \equiv 0$ and $I_t \equiv I_t^0$.

Theorem 2.2 *Assume conditions (2.13) and (2.14). Then*

$$u_t = P(\tau > t) = P(Z_t > 0) \sim L_\theta(t)t^{-(1-\theta)^+},$$

where

$$L_\theta(t) = \begin{cases} c_\theta > 0, & \theta \neq 1, \theta \geq 0, \\ c_1/\log t, & \theta = 1, \end{cases}$$

$$c_\theta = \begin{cases} B(1)/\gamma(1), & \theta > 1, \\ B(1)/\Gamma(\theta)\Gamma(1-\theta)\lim t^\theta \gamma_t, & 0 < \ \theta \leq 1, \\ (Q'(1) + p(W_1 + W_2))/b, & \theta = 0, \end{cases}$$

$B(1)$ is defined in (3.2) and W_1 and W_2 in Lemma 3.3.
If $0 < \theta \leq 1$ and only (2.13) hold, then $L_\theta(t)$ is a s.v.f. as $t \to \infty$.

Corollary. Under the conditions (2.13) for the critical migration process $\{Y_t\}$ the state zero is null-recurrent when $0 \leq \theta < 1$ and non-recurrent when $\theta > 1$. If $\theta = 1$ then the state zero is null-recurrent or non-recurrent depending on additional conditions. For example, if $\theta = 1$ and (2.14) holds, then the state zero is null-recurrent.

Comment. The case $\theta < 0$ will be considered in an other paper. Corresponding limit theorems for the both processes $\{Y_t\}$ and $\{Z_t\}$ are also obtained. An open problem is to consider the case when the emigration components η_{1t} and η_{2t} are not bounded random variables.

3 Preliminaries

The results of this section are of interest in themselves and are used in the next section.

Lemma 3.1 *The generating function of $\{u_k\}$ admits the representation*

$$U(s) = \sum_{k=0}^{\infty} u_k s^k = \frac{B(s)}{(1-s)\gamma(s)}, \tag{3.1}$$

where

$$\gamma(s) = \sum_{k=0}^{\infty} \gamma_k s^k, \qquad B(s) = \sum_{t=0}^{\infty} \gamma_t \left[1 - Q(F_t) - p\sum_{k=1}^{\infty} W(Z_{k-1}, F_t)s^k \right] s^t \tag{3.2}$$

and $Q(s)$ is defined by (2.12).

Proof. From (2.4) as $s = 0$ it follows that

$$
\begin{aligned}
u_t = 1 - \Psi(t,0) &= [1 - Q(F_t)]\gamma_t + \sum_{k=1}^{t} u_{t-k}(\gamma_{k-1} - \gamma_k) - \\
&\quad - p \sum_{k=0}^{t-1} W(Z_{t-1-k}, F_k)\gamma_k.
\end{aligned}
$$

Now multiplying by s^t and summarizing t from 0 to ∞ one obtains that

$$
\begin{aligned}
U(s) &= \sum_{t=0}^{\infty}[1 - Q(F_t)]\gamma_t s^t + U(s) \sum_{k=1}^{\infty}(\gamma_{k-1} - \gamma_k)s^k - \\
&\quad - p \sum_{k=0}^{\infty} \left(\sum_{j=1}^{\infty} W(Z_{j-1}, F_k)s^j \right) \gamma_k s^k,
\end{aligned}
$$

which proves (3.1).

Lemma 3.2 *Assume (2.13) and (2.14).*
(i) If $\theta \neq 0$, then for $0 \leq s < 1$ as $t \to \infty$

$$\gamma_t(s) \sim c_\theta(s)t^{-\theta}, \quad c_\theta(s) > 0. \tag{3.3}$$

If only (2.13) hold then $c_\theta(s)$ in (3.3) is a s.v.f. of t.
(ii) If $\theta = 0$, then $\lim_{t \to \infty} \gamma_t = c_0 > 0$ and as $s \uparrow 1$,

$$\gamma(s) = \sum_{t=0}^{\infty} \gamma_t s^t = \frac{c_0}{1-s}L\left(\frac{1}{1-s}\right) + C\log\left(\frac{1}{1-s}\right)L\left(\frac{1}{1-s}\right)(1 + o(1)), \tag{3.4}$$

where $L(x)$ is a s.v.f. and $L(x) \to 1$ as $x \to \infty$.

Proof. (i) We will use the representation

$$
\gamma_t(s) = \delta(s) \prod_{j=1}^{t-1} f(F_j(s)) \Big/ \left\{ \prod_{j=1}^{t-1} F(F_j(s)) \right\}^{N_1} \left\{ \prod_{j=1}^{t-1} F_j(s) \right\}^{N_2},
$$

where $f(s) = pE\{F^{N_1-\eta_{1t}}(s)s^{N_2-\eta_{2t}}\} + qF^{N_1}(s)s^{N_2} + rF^{N_1}(s)s^{N_2}g(s)$ is a p.g.f. for which $f(0) > 0$ and $0 < f'(1) < \infty$.
Since for every $s \in [0,1)$ there exists $k \geq 0$ such that $F_k \leq s \leq F_{k+1}$, then it is not difficult to see that $\underline{\gamma}_t(s) \leq \gamma_t(s) \leq \overline{\gamma}_t(s)$, where

$$
\underline{\gamma}_t(s) = \delta(s) \prod_{j=1}^{t-1} f(F_j(F_k)) \Big/ \left\{ \prod_{j=1}^{t-1} F_{j+1}(F_{k+1}) \right\}^{N_1} \left\{ \prod_{j=1}^{t-1} F_j(F_{k+1}) \right\}^{N_2},
$$

$$
\overline{\gamma}_t(s) = \delta(s) \prod_{j=1}^{t-1} f(F_j(F_{k+1})) \Big/ \left\{ \prod_{j=1}^{t-1} F_{j+1}(F_k) \right\}^{N_1} \left\{ \prod_{j=1}^{t-1} F_j(F_k) \right\}^{N_2}.
$$

Now applying Lemma 4 of Zubkov(1972) one obtains $\gamma_t(s) \sim c_1(s)t^{-\theta} \sim \overline{\gamma}_t(s)$ which proves (3.3).

(ii) From $\theta = 0$ it follows that $r = p(e_1 + e_2)/\lambda$ and $\delta(F_k) = 1 + pa_k/\lambda$, where

$$a_k = \lambda(H(F_{k+1}^{-1}, F_k^{-1}) - 1) - (e_1 + e_2)(1 - g(F_k)).$$

On the other hand $\underline{a}_k \le a_k \le \overline{a}_k$ where

$$\underline{a}_k = \lambda(H(F_{k+1}^{-1}, F_{k+1}^{-1}) - 1) - (e_1 + e_2)(1 - g(F_k)),$$

$$\overline{a}_k = \lambda(H(F_k^{-1}, F_k^{-1}) - 1) - (e_1 + e_2)(1 - g(F_k)).$$

Using that $1 - F(s) = 1 - s - \tilde{b}(s)(1 - s)^2$, where $\tilde{b}(s) \uparrow b$ as $s \uparrow 1$, one can obtain

$$F^{-1}(F_k) = 1 + (1 - F_k)\frac{1 - \tilde{b}(F_k)(1 - F_k)}{F_k + \tilde{b}(F_k)(1 - F_k)^2} + O((1 - F_k)^2), \quad k \to \infty.$$

Therefore, as $k \to \infty$ one can show that

$$\begin{aligned}
\underline{a}_k &= \lambda(H(F_{k+1}^{-1}, F_{k+1}^{-1}) - 1) - (e_1 + e_2)\lambda(1 - F_k) + (e_1 + e_2)[\lambda(1 - F_k) - (1 - g(F_k))] \\
&= \underline{c}_k(1 - F_k)^2(c_k + o(1)) + (e_1 + e_2)[g(F_k) - 1 + \lambda(1 - F_k)], \quad \underline{c}_k \to c < \infty.
\end{aligned}$$

From Lemma 2 of Zubkov(1972) it follows that $\sum_{k=1}^{\infty} R_1^g(1 - F_k)$ converges, where $R_1^g(1 - s) = g(s) - 1 + \lambda(1 - s)$. Hence $\sum_{k=1}^{\infty} \underline{a}_k < \infty$. Similarly one can show that $\sum_{k=1}^{\infty} \overline{a}_k < \infty$. Therefore $\prod_{k=1}^{\infty}(1 + \frac{p}{\lambda}a_k) < \infty$, which proves that $\lim_{t\to\infty} \gamma_t = c_0 < \infty$.

One can see that as $t \to \infty$

$$\begin{aligned}
\gamma_t &= \prod_{j=0}^{t-1} \delta(F_j) = c_0 \prod_{j=t}^{\infty} \delta^{-1}(F_j) = c_0 exp\left\{-\sum_{j=t}^{\infty} \log(1 + \frac{p}{\lambda}a_j)\right\} \\
&= c_0 exp\left\{-\sum_{j=t}^{\infty}(\frac{p}{\lambda}a_j + \varepsilon_j)\right\}, \quad where\ \varepsilon_j = O(a_j^2), \\
&= c_0 exp\left\{-\sum_{j=t}^{\infty}\left(c_j(1 - F_j)^2(1 + o(1)) + (e_1 + e_2)R_1^g(1 - F_j) + \varepsilon_j\right)\right\}, \quad where\ c_j \to c, \\
&= c_0 exp\left\{-\sum_{j=t}^{\infty} c_j(1 - F_j)^2(1 + o(1))\right\} exp\left\{\sum_{j=t}^{\infty}((e_1 + e_2)R_1^g(1 - F_j) + \varepsilon_j)\right\} \\
&= c_0\left(1 - \sum_{j=t}^{\infty} c_j(1 - F_j)^2\right) L(t)(1 + o(1)) \\
&= \left(c_0 L(t) - CL(t)t^{-1}\right)(1 + o(1)),
\end{aligned}$$

where $L(t)$ is a s.v.f. and $L(t) \to 1$ as $t \to \infty$.

Now, it is not difficult to obtain (3.4).

Lemma 3.3 *Under the conditions (2.13)*

$$W_1 = \sum_{t=1}^{\infty} E(\eta_{1t} + \eta_{2t} - Z_t)\mathbb{I}_{\{\eta_{1t} - Z_t \ge 0, Z_t > 0\}} < \infty,$$

$$W_2 = \sum_{t=1}^{\infty} E\mu_t(Z_t - \eta_{1t})\mathbb{I}_{\{\mu_t(Z_t - \eta_{1t}) \ge 0, Z_t > \eta_{1t}\}} < \infty.$$

Proof. For the Green's function of $\{Z_t\}$ we have

$$G(k) \;=\; \sum_{t=1}^{\infty} P(Z_t = k) = \frac{1}{r(1 - g_0(0))} \sum_{t=1}^{\infty} P(Y_{t+1} = k, \min_{1 \le i \le t+1} Y_i > 0) \quad (3.5)$$

$$= \frac{1}{r(1 - g_0(0))} \sum_{t=1}^{\infty} {}_0p_{0k}(t) < \infty, \quad k = 0, 1, 2, \ldots,$$

where ${}_0p_{0k}(t)$ are taboo-probabilities for $\{Y_t\}$ (see Chung(1960), Ch.1, Th.9.3).

Let $h_{ij} = P(\eta_{2t} = i, \eta_{1t} = j)$. Hence from (3.5) it follows that

$$W_1 \;=\; \sum_{t=1}^{\infty} \sum_{i,j=0}^{N_1,N_2} \sum_{k=1}^{i} h_{ij} P(Z_t = k)(i + j - k) \quad (3.6)$$

$$= \sum_{i,j=0}^{N_1,N_2} \sum_{k=1}^{i} h_{ij}(i + j - k) G(k) < \infty.$$

For W_2 it is not difficult to obtain

$$W_2 \;\le\; N_2 \sum_{t=1}^{\infty} P\Big(\eta_{2t} - \sum_{i=1}^{(Z_t - \eta_{1t})^+} X_{i,t} \le 0\Big) \quad (3.7)$$

$$= N_2 \sum_{t=1}^{\infty} \sum_{k=1}^{\infty} P(Z_t = k) P\Big(\eta_{2t} - \sum_{i=1}^{(k - \eta_{1t})^+} X_{i,t} \le 0\Big).$$

On the other hand,

$$_0p_{k0}(1) \;=\; qf_0^k + pP\Big(\Big(\sum_{i=1}^{(k-\eta_{1t})^+} X_{i,t} - \eta_{2t}\Big)^+ = 0\Big) \quad (3.8)$$

$$\ge pP\Big(\eta_{2t} - \sum_{i=1}^{(k-\eta_{1t})^+} X_{i,t} \le 0\Big).$$

From (3.7), using (3.5) and (3.8), it follows that

$$W_2 \;\le\; \frac{N_2}{p} \sum_{t=1}^{\infty} \sum_{k=1}^{\infty} P(Z_t = k) {}_0p_{k0}(1)$$

$$= \frac{N_2}{pr(1 - g_0(0))} \sum_{t=1}^{\infty} \sum_{k=1}^{\infty} {}_0p_{0k}(t) {}_0p_{k0}(1)$$

$$= \frac{N_2}{pr(1 - g_0(0))} \sum_{t=1}^{\infty} {}_0p_{00}(t + 1)$$

$$\le \frac{N_2}{p} G(0) < \infty.$$

4 Proof of the Theorem 2.2

Further on we will use the following relations, as $t \to \infty$:

$$\left.\begin{array}{l} 1 - F_t \sim 1/bt, \; 1 - Q(F_t) \sim Q'(1)/bt, \\[2mm] 1 - F_{t+1}^{-n} F_t^{-m} \sim -(n + m)/bt, \quad n \ge 0, \; m \ge 0. \end{array}\right\} \quad (4.1)$$

(i) Let $\theta > 1$. Then from (2.2) and (3.2) using (3.3), (4.1) and Lemma 3.3 it is not difficult to obtain $B(1) < \infty$ and $\gamma(1) < \infty$. Therefore as $s \uparrow 1$,

$$(1 - s)U(s) \to B(1)/\gamma(1) = u \geq 0. \tag{4.2}$$

On the other hand, from (1.3) and (2.4) one has for $0 < y < 1$:

$$
\begin{aligned}
u_t = 1 - \Psi(t, 0) &\geq 1 - \Psi(t, y) = 1 - \delta(y)\Psi(t - 1, 0) - pW(Z_{t-1}, y) \\
&= (1 - \delta(y))u_{t-1} + \delta(y)(1 - \Psi(t - 1, F(y))) - pW(Z_{t-1}, y) ,
\end{aligned}
$$

Iterating one obtains

$$
\begin{aligned}
u_t \geq{}& \gamma_t(y)[1 - Q(F_t(y))] + \sum_{k=0}^{t-1} u_{t-k-1}[1 - \delta(F_k(y))]\gamma_k(y) - \\
& - p\sum_{k=0}^{t-1} W(Z_{t-1-k}, F_k(y))\gamma_k(y).
\end{aligned}
$$

Therefore

$$(1 - s)U(s) \geq \frac{B(y, s)}{\sum_{t=0}^{\infty} \gamma_t(y)s^t}, \tag{4.3}$$

where

$$B(y, s) = \sum_{t=0}^{\infty} \gamma_t(y) \left\{ 1 - Q(F_t(y)) - p\sum_{k=1}^{\infty} W(Z_{k-1}, F_t(y))s^k \right\} s^t.$$

The relations (4.1) are fulfilled for $F_t = F_t(y)$, $0 < y < 1$, then similarly to (4.2) one can obtain

$$\lim_{s \uparrow 1} \frac{B(y, s)}{\sum_{t=0}^{\infty} \gamma_t(y)s^t} = u(y) > 0. \tag{4.4}$$

since $\lim_{s \uparrow 1} B(y, s) = B(y) > 0$ for $0 < y < 1$.

From (4.2)–(4.4) it follows that $u > 0$.

Now by a Tauberian theorem $u_t \downarrow u$, i.e. $P(\tau = \infty) = u > 0$ and $E\tau = U(1) = \infty$.

(ii) Let $\theta = 1$. Assume (2.13) only. The same arguments as in the case (i) show that $B(1) < \infty$. On the other hand from Lemma 3.2 and Feller(1966) Th.1, Ch.8.9 and Th.5, Ch.13.5 it follows that as $s \uparrow 1$,

$$\sum_{t=0}^{\infty} \gamma_t s^t \sim L(\frac{1}{1 - s}), \qquad \text{where } L(x) \text{ is a s.v.f.}$$

Therefore as $s \uparrow 1$,

$$U(s) \sim (1 - s)^{-1}L_1(\frac{1}{1 - s}) , \qquad \text{where} \qquad L_1(\frac{1}{1 - s}) = B(1) \Big/ L(\frac{1}{1 - s}). \tag{4.5}$$

Now since $u_t \downarrow$, applying a Tauberian theorem (see Feller(1966), Th.5, Ch.13.5) one obtains from (4.5) that $u_t \sim L_1(t)$ as $t \to \infty$. Obviously, $E\tau = \sum_{t=0}^{\infty} u_t = \infty$ and τ will be a proper r.v. iff $L_1(t) \to 0$, as $t \to \infty$.

If in addition the conditions (2.14) are fulfilled, then from Lemma 3.2 it follows that in (4.5) $L_1(1/(1 - s)) = c_1/\log(1/(1 - s))$, $c_1 > 0$. Therefore $u_t \sim c_1/\log t$ and $P(\tau < \infty) = 1$.

(iii) Let $0 < \theta < 1$. Assume (2.13) only. Similarly to the case (i) one can obtain $B(1) < \infty$ and

$$\sum_{t=0}^{\infty} \gamma_t s^t \sim (1-s)^{\theta-1} L_1(\frac{1}{1-s}) , \qquad \text{where} \qquad L_1(\frac{1}{1-s}) = \Gamma(1-\theta)L(\frac{1}{1-s}).$$

Now from (3.1) one has $U(s) \sim (1-s)^{-\theta}L_1(1/(1-s))$, $s \uparrow 1$. Hence (see the proof of (ii)) $u_t \sim t^{\theta-1}L_1(t)/\Gamma(\theta)$, as $t \to \infty$. If one additionally assumes (2.14), then by Lemma 3.2 it follows that $u_t \sim c_\theta t^{\theta-1}$, where $c_\theta = B(1)/\Gamma(\theta)\Gamma(1-\theta) \lim t^\theta \gamma_t$.

(iv) Let $\theta = 0$. In this case from (2.2) and (3.2) using Lemma 3.2 and (4.1) it is not difficult to obtain that as $s \uparrow 1$,

$$\left. \begin{array}{l} B(s) = c_0 K \log \frac{1}{(1-s)} , \quad B'(s) = \frac{c_0 K}{(1-s)} , \quad B''(s) = \frac{c_0 K}{(1-s)^2} , \\ \gamma(s) = \frac{c_0}{(1-s)}L\left(\frac{1}{1-s}\right) - c_0 \log \frac{1}{(1-s)}CL\left(\frac{1}{1-s}\right)(1+o(1)) , \\ \gamma'(s) = \frac{c_0}{(1-s)^2}L\left(\frac{1}{1-s}\right) - \frac{c_0}{(1-s)}CL\left(\frac{1}{1-s}\right)(1+o(1)) , \\ \gamma''(s) = \frac{2c_0}{(1-s)^3}L\left(\frac{1}{1-s}\right) - \frac{c_0}{(1-s)^2}CL\left(\frac{1}{1-s}\right)(1+o(1)) , \end{array} \right\} \qquad (4.6)$$

where $K = (Q'(1)+pW)/b$, $L(x)$ and $L_1(x)$ are s.v.f. and $L(x) \to 1$ as $x \to \infty$. Now from (4.6), differentiating (3.1), one can obtain

$$U'(s) = \sum_{t=1}^{\infty} tu_t s^t = o\left(\frac{1}{1-s} \log \frac{1}{1-s}\right). \qquad (4.7)$$

Differentiating $T(s) = \sum_{t=1}^{\infty} P(\tau = t)s^t = \sum_{t=1}^{\infty}(u_{t-1} - u_t)s^t = 1 - (1-s)U(s)$ and using (4.6), it is not difficult to obtain as $s \uparrow 1$,

$$\begin{array}{l} T'(s) = -\dfrac{B'(s)}{\gamma(s)} + \dfrac{B(s)\gamma'(s)}{\gamma^2(s)} \sim K \log \dfrac{1}{1-s}, \\ T''(s) = -\dfrac{B''(s)}{\gamma(s)} + 2\dfrac{B'(s)\gamma'(s)}{\gamma^2(s)} - B(s)\left(\dfrac{1}{\gamma(s)}\right)'' \sim \dfrac{K}{1-s} . \end{array}$$

By a Tauberian theorem it follows that as $t \to \infty$

$$R_t = \sum_{k=1}^{t} k^2 P(\tau = k) \sim Kt .$$

Now from Feller(1966), Th.2, Ch.8.9 one obtains

$$\frac{t^2 u_t}{R_t} \to c, \qquad 0 \le c \le \infty$$

and hence $u_t \sim cKt^{-1}$.

If one assumes $c = \infty$, then by the same theorem it follows that u_t is a s.v.f. Therefore $tu_t(\log t)^{-1} \to 0$ (see (4.7)). If $c = 0$ then R_t is a s.v.f. which is impossible. Finally $0 < c < \infty$. One can see that $c = 1$ besause of $\sum_{i=1}^{t} i^2 P(\tau = i) \sim \sum_{k=1}^{t} ku_k \sim Kt$.

References

Chung Kai Lai (1960) *Markov Chains with Stationary Transition Probabilities* Springer-Verlag, Berlin Götingen Heidelberg.

Feller W. (1966) *An Introduction to Probability Theory and its Applications, Vol 2.* Wiley, New York.

Foster, J.H. (1971) A limit theorem for a branching process with state-dependent immigration. *Ann. Math. Statist.* **42**, 1773-1776.

Grey, D.R. (1988) Supercritical branching processes with density independent catastrophes. *Math. Proc. Camb. Phil. Soc.* **104**, 413-416.

Han, L.V. (1980) Limit theorems for Galton-Watson branching processes with migration. *Siberian Math. J.* **28**, No2, 183-194 (In Russian).

Ivanoff, B.G. and Seneta, E. (1985) The critical branching process with immigration stopped at zero. *J. Appl. Prob.* **22**, 223-227.

Kaverin, S.V. (1990) A Refinement of Limit Theorems for Critical Branching Processes with an Emigration. *Theory Prob. Appl.*, **35**, 570-575.

Nagaev, S.V. and Han, L.V. (1980) Limit theorems for critical Galton-Watson branching process with migration. *Theory Prob. Appl.* **25**, 523-534.

Pakes, A.G. (1971) A branching process with a state-dependent immigration component. *Adv. Appl. Prob.* **3**, 301-314.

Seneta, E. and Tavare, S. (1983) A note on models using the branching processes with immigration stopped at zero. *J. Appl. Prob.* **20**, 11-18.

Vatutin, V.A. (1977a) A critical Galton-Watson branching process with emigration. *Theory Prob. Appl.* **22**, 482-497.

Vatutin, V.A. (1977b) A conditional limit theorem for a critical branching process with immigration. *Math.Zametki* **21**, 727-736 (In Russian).

Yanev, G.P. and Yanev, N.M. (1991) On a new model of branching migration processes.*C. R. Acad. Bulg. Sci.*, **44**, No3, 19-22.

Yanev, G.P. and Yanev, N.M. (1993) On Critical Branching Migration Processes with Predominating Emigration. *Preprint, Institute of Math., Sofia* No 1, pp.37.

Yanev, N.M. and Mitov, K.V. (1980) Controlled branching processes: The case of random migration. *C. R. Acad. Bulg. Sci.* **33**, No4, 433-435.

Yanev, N.M. and Mitov, K.V. (1981) Critical branching migration processes. *Proc. 10th Spring Conf. Union of Bulg. Math.*, 321-328 (In Russian).

Yanev, N.M. and Mitov, K.V. (1983) The life-periods of critical branching processes with random migration. *Theory Prob. Appl.*, **28**, No3, 458-467.

Yanev, N.M. and Mitov, K.V. (1985) Critical branching processes with non-homogeneous migration. *Ann. Prob.*, **13**, 923-933.

Yanev, N.M., Vatutin, V.A. and Mitov, K.V. (1986) Critical branching processes with random migration stopped at zero. *Proc. 15th Spring Conf. Union of Bulg. Math.*, 511-517 (In Russian).

Zubkov, A.M. (1972) Life-periods of a branching process with immigration. *Theory Prob. Appl.*, **17**, 179-188.

SOME RESULTS FOR MULTITYPE BELLMAN-HARRIS BRANCHING PROCESSES WITH STATE-DEPENDENT IMMIGRATION

Kosto V. Mitov[*]

Abstract

Bellman-Harris branching processes with $r>1$ types of particles allowing immigration only in the state zero are considered. The asymptotic behaviour of the probability of extinction and non-extinction in the critical case under the assumptions that the means of the immigration and the duration of the stay at zero are infinite is obtained. Limit theorems for the functional $(u,Z(t))$ are also provided.

Bellman-Harris branching processes; state-dependent immigration; limit theorems

60J80

1 Introduction.

A model of a single type Bellman-Harris branching processes with immigration in the state zero was defined by Mitov and Yanev (1985) as a generalization of the Foster (1971) and Pakes (1971) , (1975), (1978) model for Bienayme-Galton-Watson branching processes. Further, this model has been studied by Mitov and Yanev (1989) in the critical case and by Slavtchova and Yanev (1991) in the non-critical cases. Multitype analogs of these processes were considered by Mitov (1989) in the critical case. Some results in noncritical cases have been established in (Slavtchova(1991)). The present paper continues the investigation of multitype Bellman-Harris branching processes allowing immigration only in the state zero. Limit theorems are obtained in the critical case when the means of the immigration and the duration of the stay at zero are infinite. These results generalize those for the single type case.

2 Definition and functional equations.

Let $r >1$ be an integer constant. Denote by $T_1,T_2,...T_r$ the types of the particles. For r-dimensional vectors $x=(x_1,x_2,...x_r)$, $y=(y_1,y_2,...y_r)$,

---
[*] Postal address: G. Kochev str. 8, bl. 8, vh. A, apt. 16, 5800 Pleven, Bulgaria.
Supported by the National Foundation for Scientific Investigations, grant MM-4/94.

$1=(1,1,...1)$, $0=(0,0,...0)$ etc., we denote $(x,y)=(x_1y_1,x_2y_2,...x_ry_r)$, $x^y=(x_1^{y_1},x_2^{y_2},...x_r^{y_r})$ and $x{\geq}y$ or $x{>}y$ if $x_i \geq y_i$ or $x_i{>}y_i$ for $i=1,2,...r$ respectively.

Now we recall the definition from (Mitov (1989)).

Let us have on the probability space $(\Omega,A,\mathbf{P})$ three independent sets of random variables:

i) $X=\{X_i ,i=1,2,...\}$ is a set of independent identicaly distributed random variables (i.i.d.r.v.) with distribution function (d.f.) $K(t)=\mathbf{P}\{X_i \leq t\}$, $K(0+)=0;$

ii) $Y=\{\mathbf{Y}_i =(Y_{1i},Y_{2i},...Y_{ri}),\ i=1,2,...\}$ are i.i.d. random vectors with integer-valued, non-negative components and probability generating function (p.g.f) $g(s)=\mathbf{E}\{s^{Y_i}\}=\mathbf{E}\{s_1^{Y_{1i}} s_2^{Y_{2i}} ...s_r^{Y_{ri}}\}$, $g(0)=0;$

iii) $Z=\{\mathbf{Z}_{ijk}(t) = (Z_{ijk}^1(t),Z_{ijk}^2(t),...Z_{ijk}^r(t)),i \geq 1, j \geq 1, k = 1,2,...r,t \geq 0\}$ is a set of i.i.d. r-type Bellman-Harris branching processes with particle-life distribution functions $G_k(t),G_k(0+) = 0,k = 1,2,...r$, and offsprings p.g.f. $f_k(s)$, $k=1,2,..r.$

We write the vector functions $G(t)=(G_1(t),G_2(t),...G_r(t))$ and $f(s)=(f_1(s),f_2(s)...f_r(s))$ and the moments $m_{ij} = \dfrac{\partial f_i(s)}{\partial s_j}\Big|_{s=1}, b_{jk}^i = \dfrac{\partial^2 f_i(s)}{\partial s_j \partial s_k}\Big|_{s=1}$,

$$M = \|m_{ij}\|, \quad \mu_k = \int_0^\infty tdG_k(t), \quad \mu = (\mu_1,\mu_2,...\mu_r).$$

We now define the sequence of independent processes
$$\mathbf{Z}_i(t) = (Z_i^1(t),Z_i^2(t),...Z_i^r(t)),t \geq 0,i \geq 1,$$
where

$$Z_i^l(t) = \sum_{k=1}^{r}\sum_{j=1}^{Y_i^k} Z_{ijk}^l(t)$$

is the number of particles of type T_l in the process $\mathbf{Z}_i(t),t \geq 0$, which starts with the random number $\mathbf{Y}_i = (Y_i^1,Y_i^2,...Y_i^r)$ of particles at the moment $t=0$.

Denote by τ_i the life-period of the process $\mathbf{Z}_i(t),t \geq 0$, i.e. τ_i is a random variable such that $\{\mathbf{Z}_i(0) = \mathbf{Y}_i \neq 0, \mathbf{Z}_i(t) \neq 0,t \in(0,\tau_i), \mathbf{Z}_i(\tau_i) = 0$.

The sequence $\eta_i = \tau_i + X_i, i = 1,2,....$ of i.i.d. r.v. defines a renewal process $\{S_i\}_{i=0}^{\infty}$ with

$$S_0 = 0,S_{i+1} = S_i +\eta_{i+1},i \geq 0, \qquad (2.1)$$

and, as usual, $N(t) = \max\{i:S_i \leq t < S_{i+1}\}$ is the number of renewal events in the interval $[0,t]$.

Now the Bellman-Harris branching process with r types of particles and immigration only in the state zero is defined by:

$$Z(t) = \begin{cases} Z_{N(t)}(t - S_{N(t)} - X_{N(t)+1}), & t - S_{N(t)} - X_{N(t)+1} \geq 0, \\ 0, & t - S_{N(t)} - X_{N(t)+1} < 0. \end{cases} \tag{2.2}$$

We shall also use the following notation: $F_k(t,s) = \mathbf{E}\{s^{Z_{ijk}(t)}\}$ is the probability generating function of the usual Bellman-Harris processes starting with a particle of type T_k, $k=1,2,...r$, $F(t,s)=(F_1(t,s),F_2(t,s),...F_r(t,s))$, $\Im(t,s)=\mathbf{E}\{s^{Z_i(t)}\} = \mathbf{E}\{s_1^{Z_i^1(t)} s_2^{Z_i^2(t)}...s_r^{Z_i^r(t)}\}= g(F(t,s))$ is the probability generating function of the process $Z_i(t), t \geq 0$, $\Phi(t,s)=\mathbf{E}\{s^{Z(t)}\}$, $Q(t,s)=1-\Im(t,s)$, $Q(t)=Q(t,0)$, $R(t,s)=1-\Phi(t,s)$, $R(t)=R(t,0)$.

It is well-known that the functions $F_k(t,s)$ satisfy the following system of integral equations (see e.g. Sevastyanov (1971)):

$$F_k(t,\mathrm{s}) = s_k(1 - G_k(t)) + \int_0^t f_k(F(t - y,s))dG_k(y) \tag{2.3}$$

$$F_k(0,\mathrm{s}) = s_k, k = 1,2,...r.$$

It is clear that $\mathbf{P}\{\tau_i \leq t\} = \mathbf{P}\{Z_i(t) = 0\}= \Im(t,0)$ and

$$D(t) = \mathbf{P}\{\eta_i \leq t\} = \mathbf{P}\{X_i + \tau_i \leq t\} = \int_0^t \Im(t-y,0)dK(y), \qquad D(0+)=0. \tag{2.4}$$

It is not difficult to show that the p.g.f. $\Phi(t,s)$ satisfies the integral equation :

$$\Phi(t,s) = 1 - K(t) - D(t) + \int_0^t \Im(t - y,s)dK(y) + \int_0^t \Phi(t - y,s)dD(y). \tag{2.5}$$

The proof is quite similar to that for the single type case and is omitted (See Mitov and Yanev (1985)).

Now we shall consider the process $Z(t)$, $t \geq 0$ under the following three conditions for regularity of critical Bellman-Harris branching processes:

(1) $0 < m_{ij} < \infty$, and the matrix $M = \|m_{ij}\|$ has the Peron root $\rho=1$ (critical case) and the right and left eigenvectors $u = (u_1, u_2,...u_r)$ and $v = (v_1, v_2,...v_r)$ corresponding to the Peron root are such that $Mu = u$, $vM = v$, $(u,v) = 1$, $(u,1) = 1$;

(2) $\qquad b_{jk}^i < \infty, \quad 0 < \dfrac{1}{2}\sum_{i,j,k} v_i b_{jk}^i u_j u_k < \infty;$

(3) $\qquad t^2(1 - G_k(t)) \to 0, \quad t \to \infty, \quad k = 1,2,...r,$

$$\mu = (\mu_1,\mu_2,...\mu_r) < \infty, \quad \text{and} \quad G_k(t) \quad \text{are non-lattice d.f.;}$$

For the moments of the size of immigration and the duration of the stay at zero we assume that some of the following conditions hold:

(A) $\quad 1 - g(s) = (a,1-s)^p, \quad 0 < p < 1, \quad v = \int_0^\infty t \, dK(t) < \infty;$

(B) $1 - g(s) = (a,1-s)^p, 0 < p < 1, 1 - K(t) = t^{-q} L(t), t \to \infty, 0 < q < 1;$

(C)
$$
\begin{cases}
n_j = \dfrac{\partial g(s)}{\partial s_j}\Big|_{s=1} < \infty, \; n = (n_1,n_2,...n_r), \\[2mm]
1 - K(t) = t^{-q} L(t), t \to \infty, 0 < q < 1;
\end{cases}
$$

(D)   There exists a density $k(t)=K'(t)$ such that $k(t)$ is directly Riemann integrable and $k(t)=O(1/t)$, $t\to\infty$.

The vector $a = (a_1,a_2,...a_r)$ is such that $(a,1)\le 1$, $a\ge 0$ and $a\ne 0$, and $L(t)$ is a function which is slowly varying at infinity.

3 Preliminary results.

Let us denote the set $\wp=\{ s = (s_1,s_2,...s_r): \; s = 1 - cu; \; c \in [0,1]\}.$

Lemma 3.1 *Under conditions (1)-(3) (see Goldstein (1978))*

$$\lim_{t\to\infty}\left|\frac{bt}{(u,v\mu)} + \frac{1}{(v,1-s)}\right|(1 - F_k(t,s)) = u_k, \quad k = 1,2,...r, \tag{3.1}$$

uniformly for $s\in \wp$ and

$$\lim_{t\to\infty}\frac{bt}{(u,v\mu)}(1 - F_k(t,0)) = u_k, \quad k = 1,2,...r. \tag{3.2}$$

Lemma 3.2 *Let (1)-(3) hold. If in addition (A) or (B) hold, then*

$$\lim_{t\to\infty}\left|\frac{bt}{(u,v\mu)} + \frac{1}{(v,1-s)}\right|^p Q(t,s) = (a,u)^p, \tag{3.3}$$

uniformly for $s\in \wp$, and

$$\left(\frac{bt}{(u,v\mu)}\right)^p Q(t) \to (a,u)^p, \quad t \to \infty. \tag{3.4}$$

Proof. From $Q(t,s) = 1 - g(F(t,s)) = (a, 1 - F(t,s))^p$ it is not difficult to obtain that

$$Q(t,s)\left|\frac{bt}{(u,v\mu)} + \frac{1}{(v,1-s)}\right|^p = \left(a, \left|\frac{bt}{(u,v\mu)} + \frac{1}{(v,1-s)}\right|(1 - F(t,s))\right)^p.$$

Now using (3.1) we obtain (3.3). Similarly (3.4) follows from (3.2).

Lemma 3.3 *Let (1)-(3) and (C) hold. Then*

$$\lim_{t\to\infty}\left|\frac{bt}{(u,v\mu)} + \frac{1}{(v,1-s)}\right|Q(t,s) = (n,u) \tag{3.5}$$

uniformly for $s \in \wp$, and

$$\left(\frac{bt}{(u,v\mu)}\right)Q(t) \to (n,u), \quad t \to \infty. \tag{3.6}$$

The proof of this lemma can be found in (Mitov(1989)).

Lemma 3.4 *i) Let the conditions (1)-(3) and (A) hold. Then*

$$\mathbf{P}\{\eta_i > t\} = 1 - D(t) \sim \left[\frac{(a,u)(u,v\mu)}{bt}\right]^p, \quad t \to \infty. \tag{3.7}$$

ii) Let the conditions (1)-(3) and (B) hold. Then
a) if $p<q$, then

$$\mathbf{P}\{\eta_i > t\} = 1 - D(t) \sim \left[\frac{(a,u)(u,v\mu)}{bt}\right]^p, \quad t \to \infty; \tag{3.8}$$

b) if $p>q$, then

$$\mathbf{P}\{\eta_i > t\} = 1 - D(t) \sim t^{-q}L(t), \quad t \to \infty; \tag{3.9}$$

c) if $p=q$, then

$$\mathbf{P}\{\eta_i > t\} = 1 - D(t) \sim t^{-p}\left\{\left[\frac{(a,u)(u,v\mu)}{b}\right]^p + L(t)\right\}, \quad t \to \infty. \tag{3.10}$$

iii) Let the conditions (1)-(3) and (C) hold. Then

$$\mathbf{P}\{\eta_i > t\} = 1 - D(t) \sim t^{-q}L(t), \quad t \to \infty. \tag{3.11}$$

The proof follows from the relation
$$1 - D(t) = \mathbf{P}\{X_i + \tau_i > t\} \sim \mathbf{P}\{X_i > t\} + \mathbf{P}\{\tau_i > t\} = 1 - K(t) + Q(t),$$
(see Feller(1971), SectionVIII.8, (8.15) and (8.16)) using (3.4) and (3.6).
From the equation (2.5) we obtain

$$R(t,s) = \int_0^t R(t-y,s)dD(y) + \int_0^t Q(t-y,s)dK(y), \qquad (3.12)$$

which is an equation of the renewal type and for each $s \in [0,1]^{\times r}$ has a unique solution

$$R(t,s) = \int_0^t W(t-y,s)dH(y), \qquad (3.13)$$

where $H(t) = \sum_{n=0}^{\infty} D^{*n}(t)$ is the renewal function for the process, defined by (2.1) and

$$W(t,s) = \int_0^t Q(t-y,s)dK(y). \qquad (3.14)$$

If there exists a density $k(t)=K'(t)$, it is not difficult to show that

$$R(t,s) = \int_0^t Q(t-y,s)X(y)dy \qquad (3.15)$$

where $X(t)$ is the unique solution of the renewal equation

$$X(t) = k(t) + \int_0^t X(t-y)dK(y). \qquad (3.16)$$

For $X(t)$ we obtain the following results.

Lemma 3.5 *Assume that conditions (1)-(3) and (D) hold and $1/2<p,q<1$. Then:*

i) *If additionaly (A) or (B) with $p<q$ holds, then*

$$X(t) \sim \frac{\sin \pi p}{\pi}\left[\frac{b}{(a,u)(u,v\mu)}\right]^p t^{p-1}, t \to \infty. \qquad (3.17)$$

ii) *If additionaly (C) or (B) with $p>q$ holds, then*

$$X(t) \sim \frac{\sin \pi q}{\pi}\frac{t^{q-1}}{L(t)}, t \to \infty. \qquad (3.18)$$

iii) *If additionaly (B) with $p=q$ holds, then*

$$X(t) \sim t^{p-1}\frac{\sin \pi p}{\pi}\Big/\left\{\left[\frac{(a,u)(u,v\mu)}{b}\right]^p + L(t)\right\}, t \to \infty. \qquad (3.19)$$

The proof of this lemma is similar to the one in the single type case (see Mitov and Yanev (1989)). We use (3.8)-(3.11) instead of the coresponding result for one type processes.

4 Probability of extinction.

In this section we consider the probability of extinction or non-extinction of the process $Z(t)$, $t \geq 0$. Basic tools for the investigation are the equations (2.5) and (3.15) with $s=0$.

Theorem 4.1 *Let (1)-(3) hold.*

i) If additionaly (A) and $1/2 < p < 1$, then

$$P\{Z(t) = 0\} \sim v \left[\frac{b}{(a,u)(u,v\mu)} \right]^p \frac{\sin \pi p}{\pi} t^{p-1} / L(t), t \to \infty. \tag{4.1}$$

ii) If additionaly (C) and $1/2 < q < 1$, then

$$P\{Z(t) = 0\} \to 1, \quad t \to \infty. \tag{4.2}$$

iii) If additionaly (C), (D) and $1/2 < q < 1$, then

$$P\{Z(t) \neq 0\} \sim (n,u) \left[\frac{(a,u)(u,v\mu)}{b} \right]^q \frac{\sin \pi q}{\pi} t^{q-1} \log t, t \to \infty. \tag{4.3}$$

Proof. *i)* Under assumptions of *i)*, (3.7) holds and $v = \int_0^\infty t dK(t) < \infty$. Now applying Theorem 3 of Erickson (1970) to (2.5) with $s=0$ we get

$$\Phi(t,0) \sim C(p) \left(\int_0^\infty 1 - K(t)dt \right) \Big/ \left(\int_0^t 1 - D(y)dy \right),$$

where $C(p)=(\sin\pi p)/(\pi(1-p))$. From (3.7) it follows that $\int_0^t 1 - D(y)dy \sim tD(t)/(1-q)$, $t \to \infty$, (see Feller (1971), Section VIII.9, Theorem 1), which implies (4.1).

ii) In this case $1-K(t)$ has the same asymptotics as $1-D(t)$. The conditions in this case allow us to use a renewal theorem (Theorem 1 of Anderson and Athreya (1987)), which gives

$$\Phi(t,0) \sim C(q) \left(\int_0^t 1 - K(y)dy \right) \Big/ \left(\int_0^t 1 - D(y)dy \right), \quad t \to \infty,$$

where $C(q)=[(2-q)B(1+q-q,2-q)]^{-1}=1$. Applying Theorem 1 (Feller (1971), VIII.9) to (3.11) we obtain (4.2).

iii) To prove (4.3) we shall use (3.15) with $s=0$. Let c be a constant such that $c \in (0,1)$. Then

$$R(t) = \int_0^{tc} Q(t - y)X(y)dy + \int_{tc}^t Q(t - y)X(y)dy = I_1(t) + I_2(t).$$

Using (3.6) and (3.18) and integral properties of regularly varying functions (see Seneta (1976), Section 1.5), we obtain that

$$I_2(t) = \int_{tc}^t Q(t - y)X(y)dy \sim X(t) \int_0^{t(1-c)} Q(y)dy \sim X(t)\frac{(u,n)(u,v\mu)}{b} \log t, t \to \infty.$$

Finally, it is not difficult to show that $I_1(t)=o(I_2(t))$, which completes the proof of (4.3) and the theorem.

Theorem 4.2 *Let (1)-(3) and (B) hold.*
i) If $p<q$, then

$$\mathbf{P}\{Z(t) = 0\} \sim \left[\frac{b}{(a,u)(u,v\mu)}\right]^p \frac{\sin \pi p}{\pi} B(p,1-q)t^{p-q} L(t), \, t \to \infty. \quad (4.4)$$

ii) If $p=q$, then

$$\mathbf{P}\{Z(t) = 0\} \sim \left\{1 + \frac{1}{L(t)}\left[\frac{(a,u)(u,v\mu)}{b}\right]^p\right\}^{-1}, \, t \to \infty. \quad (4.5)$$

iii) If $p>q$, then

$$\mathbf{P}\{Z(t) = 0\} \to 1, \quad t \to \infty. \quad (4.6)$$

Proof. In all three cases we can apply a renewal theorem (see Anderson and Athreya (1987), Theorem 1) to (2.5) with $s=0$ which gives us

$$\Phi(t,0) \sim \frac{\int_0^t (1 - K(y))dy}{\int_0^t (1 - D(y))dy} \frac{1}{(2-d)B(1+c-d,2-c)}, \quad t \to \infty,$$

where $B(.,.)$ is the beta function d is the degree of $1-K(t)$, and c is the degree of $1-D(t)$. Thus in the case *i)* $d=p$ and $c=q$; in the case *ii)* $d=c=p=q$ and finally in the case *iii)* $d=c=q$. Now (4.4), (4.5) and (4.6) follow immediately.

Theorem 4.3 *Let (1)-(3), (B) and (D) hold and $1/2<p,q<1$.*
i) If $p>q$, then

$$\mathbf{P}\{Z(t) \neq 0\} \sim \frac{t^{q-p}}{L(t)}\left[\frac{(a,u)(u,v\mu)}{b}\right]^p \frac{\sin \pi q}{\pi} B(q,1-p), \quad t \to \infty. \quad (4.7)$$

ii) If $p=q$, then

$$\mathbf{P}\{Z(t) \neq 0\} \sim \left\{1 + \left[\frac{b}{(a,u)(u,v\mu)}\right]^p L(t)\right\}^{-1}, \, t \to \infty. \quad (4.8)$$

Proof. For some fixed constant $T>0$ from (3.15) with $s=0$ we can write

$$R(t) = \int_0^T Q(t-y)X(y)dy + \int_T^{t-T} Q(t-y)X(y)dy + \int_{t-T}^t Q(t-y)X(y)dy = \quad (4.9)$$
$$= I_1(t) + I_2(t) + I_3(t),$$

for t large enough.

We have the following estimations of these three integrals:

Setting $y=xt$ in $I_2(t)$ and using Theorem 2.1 of Seneta (1976),Section 2.1, we obtain

$$I_2(t) = \int_T^{t-T} Q(t-y)X(y)dy \sim Q(t)X(t)\int_0^1 (1-x)^{-p}x^{1-q}dx, \quad t \to \infty. \quad (4.10)$$

We estimate $I_1(t)$ via

$$I_1(t) = \int_0^T Q(t-y)X(y)dy \le Q(t-T)\int_0^T X(y)dy = O(t^{-p}) = o(t^{q-p}), t \to \infty, \quad (4.11)$$

because $Q(t)$ is decreasing function of $t \to \infty$ and $Q(t)=O(t^{-p})$.

Using a property of the regularly varying functions (see Seneta (1976), Section 1.5) we obtain

$$I_3(t) = \int_{t-T}^t Q(t-y)X(y)dy \le \{ \sup_{t-T\le y<\infty} X(y)\}\int_0^T Q(y)dy \sim X(t)\int_0^T Q(y)dy, \quad (4.12)$$

as $t\to\infty$.

Now using (3.4), (3.18), (3.19) and (4.9)-(4.12) we can complete the proof.

5 Limit theorems.

In this section we consider the limit behaviour of the functional $(u,Z(t))$, when $t\to\infty$ and obtain limit theorems which are conditional on $\{Z(t)\ne0\}$ or non-conditional. It is interesting that the limit distributions of the total number of particles of any type at the moment t weighted by the components of the eigenvector u are the same as the limit distribution obtained for the processes with one type of particle. (See Mitov and Yanev (1989)). The proofs of the next theorems are similar to those ones for single type processes and we omit them. It is necessary to mention only the fact that the vectors

$$(1-e^{-\frac{\lambda u}{t}}) = (1-e^{-\frac{\lambda u_1}{t}}, 1-e^{-\frac{\lambda u_2}{t}}, ... 1-e^{-\frac{\lambda u_k}{t}})$$

and

$$(1-e^{-\frac{\lambda u}{t^x}}) = (1-e^{-\frac{\lambda u_1}{t^x}}, 1-e^{-\frac{\lambda u_2}{t^x}}, ... 1-e^{-\frac{\lambda u_k}{t^x}})$$

belong to $\wp$ for fixed $\lambda>0$, $0<x<1$ and t large enough, because of

$$1-e^{-\frac{\lambda u_k}{t}} = \frac{\lambda u_k}{t}(1+o(1)), \qquad 1-e^{-\frac{\lambda u_k}{t^x}} = \frac{\lambda u_k}{t^x}(1+o(1)),$$

as $t\to\infty$ for $\lambda>0$ and $0<x<1$. This fact allows us to use (3.3) and (3.6) in the proofs.

Theorem 5.1. *Let (1)-(3) and (D) hold. If additionaly we assume (A) or (B) with $1/2<p<q<1$ then*

$$\lim_{t\to\infty} P\left\{\frac{(u,Z(t))}{t} \le x\right\} = A(x), \quad (5.1)$$

where the distribution function $A(x)$ has Laplace transform

$$\hat{A}(\lambda) = 1 - B_{1/(1+(u,v\mu)/b\lambda)}(p,1-p) / B(p,1-p).$$

Here $B_x(.,.)$ for $0<x<1$ is the incomplete beta function.

Theorem 5.2 *Let (1)-(3), (C) and (D) hold and 1/2<q<1. Then*

$$\lim_{t\to\infty} P\left\{\frac{\log(\boldsymbol{u},\boldsymbol{Z}(t))}{\log t} \le x \Big| \boldsymbol{Z}(t) \ne \boldsymbol{0}\right\} = x, \quad 0 < x < 1. \tag{5.2}$$

Theorem 5.3 *Let (1)-(3), (D) and (B) with 1/2<q≤p<1 hold. Then*

$$\lim_{t\to\infty} P\left\{\frac{(\boldsymbol{u},\boldsymbol{Z}(t))}{t} \le x \Big| \boldsymbol{Z}(t) \ne \boldsymbol{0}\right\} = A(x), \tag{5.3}$$

where the distribution function A(x) has Laplace transform

$$\hat{A}(\lambda) = 1 - \left\{\lambda \Big/ (\lambda + \frac{(\boldsymbol{u},\boldsymbol{\nu}\mu)}{b})\right\}^{p-q} B_{\left\{\lambda \Big/ (\lambda + \frac{(\boldsymbol{u},\boldsymbol{\nu}\mu)}{b})\right\}}(q,1-p) / B(q,1-p), \quad \lambda > 0.$$

It would be interesting to obtain the multidimensional limit distribution for the process $\boldsymbol{Z}(t)$, $t>0$ when the second moments of the offspring distributions are infinite.

Acknowledgement

The author is greateful to Dr. M. Slavtchova-Bojkova and Dr. N. M. Yanev for helpful comments.

References

Anderson, K. and Athreya, K. (1987) A renewal theorem in the infinite mean case. *Ann. Appl. Prob.* **15**, 388-393.

Erickson, K. B. (1971) Strong renewal theorem with infinite mean. *Trans. Amer. Math. Soc.* **151**, 263-291.

Feller, W. (1971) *An Introduction to Probability Theory and its Applications*, Vol. 2, 2nd edn. Wiley, New York.

Foster, J. H. (1971) A limit theorem for a branching process with state-dependent immigration. *Ann. Math. Statist.* **42**, 1773-1776.

Goldstein, M. I. (1978) A uniform limit theorem and exponential limit law for critical multitype age-dependent branching processes. *J. Appl. Prob.* **15**, 235-242.

Mitov, K. V. and Yanev, N. M. (1985) Bellman-Harris branching processes with state-dependent immigration. *J. Appl. Prob.* **22**, 757-765.

Mitov, K. V. and Yanev, N. M. (1989) Bellman-Harris branching processes with a special type of state-dependent immigration. *Adv. Appl. Prob.* **21**, 270-283.

Mitov, K. V. (1989) Multitype Bellman-Harris branching processes with state-dependent immigration. *Proc. 18th Spring Conf. Union of Bulgarian Mathematicians. 423-428.* (In Bulgarian)

Pakes, A. G. (1971) A branching process with a state-dependent immigration component. *Adv. Appl. Prob.* **3**, 301-314.

Pakes, A. G. (1975) Some results for non-supercritical Galton-Watson processes with immigration. *Math. Biosci.* **42**, 71-92.

Pakes, A. G. (1978) On age distribution of a Markov chain. *J. Appl. Prob.* **15**, 65-77

Seneta, E. (1976) *Regularly Varying Functions.* Lecture Notes in Mathematics **508**, Springer-Verlag, Berlin.

Sevastyanov, B. A. (1971) *Branching Processes* (In Russian). Nauka, Moscow.

Slavtchova, M. and Yanev, N. M. (1991) Non-critical Bellman-Harris branching processes with state dependent immigration. *Serdica,* **19**, 67-79.

Slavtchova, M. (1991) Limit theorems for multitype Bellman-Harris branching processes with state dependent immigration. *Serdica,* **19**, 144-156.

BRANCHING PROCESSES AS SUMS OF DEPENDENT RANDOM VARIABLES

I. Rahimov

Romanovski Math. Inst. Tashkent and METU, Ankara, Turkey

ABSTRACT

It is well-known that branching processes can be represented as a sum of random number of independent random variables. This relation serves as a starting point for investigation of the branching process. However, characteristics of the population representing as sums of dependent terms are not usually amenable to investigation by the traditional methods in the theory of branching processes. If we use some results in the theory of summation of dependent variables:

(a) it is possible to study some new characteristics of population in the branching process without immigration (the number of r-tuples at time t_1 that have at time $t_2 > t_1$ nonempty offspring sets differing from each other in number by at most d; the number of pairs of particles having the same number of offspring at time t; reduced branching processes and so on).

(b) it is possible to study branching processes with immigration, depending on reproduction.

Key words: Discrete-time, immigrant, (k,i)-process, reproduction dependent immigration, stopping time, dependent indicators.

Let $M=\{\mu_{ki}(n),k,i\in N\},N=\{1,2,\ldots\}$,be a family of discrete-time branching processes and $f_{ki}(S)$, $|S|\leq 1$, be the generating functions of their direct offspring. If we denote by $\eta(k,n)$ the number of immigrating particles at time k and denote by $Z(n)$ the process with immigration, then it is known

Postal address: Uzbekistan, 700180, Tashkent, Yunusabad, 15-31-74.

that the process $Z(n)$, $n \in N_0 = \{0\} \cup N$, can be represented in the form of

$$Z(n) = \sum_{k=1}^{n} \sum_{i=1}^{\eta(k,n)} \mu_{ki}(n-k), \quad Z(0) = 0, \tag{1}$$

where $\mu_{ki}(n)$ is the process generating by the ith particle immigrating at time k.

Branching processes with immigration have been studied widely in the literature. In the first papers related to branching processes with immigration it was assumed that the immigration process is either homogeneous Poisson or a partial sum of independent and identically distributed random variables (Sevastyanov (1957), Foster (1969), Seneta (1970)). In source of 1970's more general models of branching processes with immigration were considered by Foster and Williamson (1971), Durham (1971), S. Nagayev (1975). Immigration processes in these papers are more general than processes with independent increments. Later on, various proofs of limit theorems for branching processes with immigration were suggested by Athreya, Parthasarathy and Sankaranarayanan (1974), Shurenkov (1976), Asmussen and Hering (1976), Badalbayev and Zubkov (1983), S. Nagayev and Asadullin (1985), I. Rahimov (1986) and others.

However, assumption of independence of immigration and reproduction processes still characterize these efforts. In our model this assumption means that the family of independent and identically distributed branching processes M and the random variables $\eta(k,n)$ are independent. Under this assumption the relation (1) serves as a starting point for the investigation of the process $Z(n)$. In this case the study of $Z(n)$ reduces to the analyses of the relation

$$H(n,S) = \prod_{k=1}^{n} h_n(k, F(n-k,S)) \tag{2}$$

for the generating function $H(n,S)$ of $Z(n)$, where $h_n(k,S) = ES^{\eta(k,n)}$, $F(n,S) = ES^{\mu_{ki}(n)}$.

However, it is impossible to get an explicitly expression for the generating function of the process with immigration if we consider a more general immigration process, depending on reproduction.

Thus a branching process from (1) corresponds to each pair $(k,i) \in N^2$. We denote the particles from (k,i)-processes by finite ordered tuples of positive integers. The tuple $(0)=N^0$ is associated with the initial particle (immigrant), and the tuple $x'=(i_1, \ldots i_k, j)$ is associated with the jth direct offspring of the particle $x = =(i_1, \ldots i_k)$. Then, corresponding to the particles in generations $1, 2, \ldots, n$ are elements of the set $I_n = \overset{n}{U} N^i$, $N^k = N^{k-1} \times N^1$, $N^1 = N$.

Let w_x^{ki} be the number of direct offspring of the particle x from the (k,i)-process and let

$$\mathcal{F}_{ki}(n) = \sigma(w_x^{ki}, x \in I_{n-k})$$

be the σ-algebra generated by evolution of the (k,i)-processes up to time n. We assume that for any j the numbers of immigrating particles $\eta(k,n)$ satisfy the condition

$$\{\eta(k,n) \leq j\} \in F_{kj}(n) = \prod_{l=1}^{k-1} \prod_{i=1}^{\eta(1,n)} \mathcal{F}_{1i}(n) \times \prod_{i=1}^{j} \mathcal{F}_{ki}(n) \times \mathcal{F}_0 , \qquad (3)$$

where $\mathcal{F}_0$ is some σ-algebra and the direct product of the random number of σ-algebras we shall understand as

$$\prod_{i=1}^{\eta} \mathcal{F}_i = \{A: A \cap \{\eta=j\} \in \prod_{i=1}^{j} \mathcal{F}_i\}, \quad \prod_{i=1}^{0} \mathcal{F}_i = \{\emptyset, \Omega\}.$$

Under condition (3), the number of immigrating particles at time k may depend on evolution of the processes generated by the particles which immigrated up to time k.

In order to study the process $Z(n)$ under the condition (3) we shall use a representation of this process as a sum of dependent random variables. For this purpose, we choose a sequence of integers $\{k_n, n \in N\}$ such that

$$P\{ \max_{1 \le k \le n} \eta(k,n) > k_n \} \to 0 \qquad (4)$$

as $n \to \infty$. If we put

$$\nu_j(n) = \chi\{i \le \eta(k,n)\}, \quad X_j(n) = \mu_{ki}(n-k)$$

for $j=(k-1)k_n+i$, ($\chi(A)$ means the indicator function of an event A), we have that

$$T_n Z(n) = T_n V_n , \qquad (5)$$

where

$$T_n = \chi\{ \max_{1 \le k \le n} \eta(k,n) \le k_n \}, \quad V_n = \sum_{j=1}^{nk_n} \nu_j(n) X_j(n) .$$

Since

$$P\{T_n = 0\} \to 0, \quad n \to \infty,$$

by the relation (4), and

$$\left| P\{\xi \chi < x\} - P\{\xi < x\} \right| \le P\{\chi = 0\} \qquad (6)$$

for any indicator χ and for an arbitrary random variable ξ, it follows from (5) that the stochastic process $Z(n)$ and the sum V_n will have the same limit distribution as $n \to \infty$.

Now we consider the sum V_n of dependent random variables $\nu_i(n) X_i(n)$. It is clear that for $j = (k-1)k_n+i$ the random variable $X_j(n)$ is measurable with respect to $F_{ki}(n)$. It follows from the condition (3) that

$$\{\eta(k,n) \le i-1\} \in F_{ki-1}(n) \Rightarrow \{\eta(k,n) \ge i\} \in F_{ki-1}(n) .$$

This means that $\nu_j(n)$ with $j=(k-1)k_n+i$ is measurable with respect to $F_{ki-1}(n)$. Therefore limit theorems obtained for sums of dependent random variables are applicable to this sum.For example, if we apply Theorem 1.3 of Eagleson (1975) to the sum V_n in (5), we have the following limit theorem for branching process $Z(n)$ with immigration, depending on reproduction.

Denote $\mu(n) = \mu_{11}(n)$,

$$\rho_n(y) = nP\{ \frac{\mu(n)}{n} > y \} \ , \ T_n(x) = n^{-1} \sum_{i=1}^{[nx]} \eta(i,n).$$

Let for some $B \in (0,\infty)$

$$\rho_n(y) \to B^{-1} e^{-y/B} \tag{7}$$

as $n \to \infty$ for any $y \in [0,\infty)$, and

$$T_n(x) \overset{P}{\to} T(x) \tag{8}$$

for $0 \le x \le 1$, where $T(x)$ is some random process stochastically continuous for $x=1$ with non-decreasing trajectory, $T(0)=0$ and $T(1)<\infty$ almost everywhere. The condition (7) is satisfied, for example, by a critical Galton-Watson processes with finite variance.

Theorem 1. If conditions (3), (7) and (8) are satisfied, then $Z(n)/n \overset{d}{\to} Z$ as $n \to \infty$, where

$$Ee^{itZ} = Eexp\{ \int_0^\infty (e^{itx}-1)dF(x) \} \ ,$$

and the function $F(x)$ is defined by the following relation:

$$F(x) = \begin{cases} \int_0^1 exp\{ - \dfrac{x}{B(1-u)} \} \dfrac{dT(u)}{B(1-u)} & , \ x>0, \\ \\ 0 & , \ x \le 0 \ . \end{cases}$$

Now we consider a corollary of Theorem 1. Let $\eta(k,n)$ be a stopping time with respect to the system $\prod_{j=1}^{i} \mathcal{F}_{kj}(n)$ for any k. Then condition (3) is

satisfied and variable $\eta(k,n)$ and processes $\{ \mu_{ij}(n) ,(i,j)\in N^2, i\neq k \}$ are

independent. Denote $f(S)=ES^{\mu_{ki}(1)}$.

Corollary 1. If $f'(1)=1$, $B=f''(1) / 2 \in (0,\infty)$ and $\alpha=E\eta(k,n)\in(0,\infty)$, then

$2Z(n)/Bn \overset{d}{\to} Z$ as $n\to\infty$, where

$$Ee^{itZ} = (1+it)^{-\alpha/B} .$$

Corollary 1 shows that the well-known theorem on convergence to the gamma-distribution is valid, without additional conditions, when the number of immigrating particles is a stopping time with respect to the system of σ-algebras generated by lives of these particles and by lives of their offspring.

The next representation which we shall consider is connected to some new characteristics of population in the branching process without immigration. Let $\mu(n)=\mu_{11}(n)$ be the discrete-time branching process. If we use the notation introduced above, then corresponding to the population at time n is a set $A_n \in E$, where E is the class of all finite subsets of $\overset{n}{\underset{1}{U}} N^i$. The process $\mu(n)=|A_n|$, $|A|$ being the number of elements of a set A, is representable in the form

$$\mu(n) = \sum_{x \in A_m} \mu^{(x)}(n-m), \quad m<n , \qquad (9)$$

where $\mu^{(x)}(n)$ is the branching process generated by a particle x alive at time m.

Let $\mathcal{B}(m,n) = \{ x\in A_m: \mu^{(x)}(n-m)\neq 0\}$ be the set of particles in A_m having a nonempty set of offspring at time n. We denote random variables $\rho(x,y)$ on $\mathcal{B}(m,n)$ by the relation

$$\rho(x,y)=\left|\mu^{(x)}(n-m)-\mu^{(y)}(n-m)\right| \ .$$

Let $d \in N_0 = N \cup \{0\}$ and let $\chi(A)$ be the indicator function of a region A as before; for arbitrary moments of time m and n with $m \le n$ we denote

$$Z_r(m,n,d) = \frac{1}{r!} \sum \xi(x_{i_1}, \dots x_{i_r}) \ , \qquad (10)$$

where the summation with respect to $x_{i_1}, \dots, x_{i_r} \in \mathcal{B}(m,n)$, $r \ge 2$ and

$$\xi(x_{i_1}, \dots, x_{i_r}) = \prod_{1 \le i < j \le r} \chi(x_i \ne x_j, \ \rho(x_i, x_j) \le d)$$

Obviously, $Z_r(m,n,d)$ is equal to the number of r-tuples of particles in A_m that have at time n nonempty offspring sets differing from each other in number by at most d. In particular, $Z_2(m,n,0)$ is the number of pairs of particles in $\mathcal{B}(m,n)$ having the same number of offspring at time n (Rahimov (1987)).

Since the number of elements of $\mathcal{B}(m,n)$ is a random variable, the $Z_r(m.n,d)$ is a random sum of dependent indicators. Quite a few articles have been devoted to the investigation of sums of the form (10) with a deterministic number of summands. For instance, if we use results of Sevastyanov (1972), Ambrosimov (1976), Silverman and Brown (1978), we can obtain limit theorems for $Z_r(m,n,d)$ and for similar characteristics of branching processes.

Now we shall present a limit theorem for $Z_2(m,n,d)$. It will be assumed below that $\mu(n)$ is a nonperiodic discrete time Markov branching process, $\mu(0)=1$ and

$$A(m,n,d)=E[Z_2(m,n,d)\mid \mu(m)>0] \ , \quad B=E\mu(1)(\mu(1)-1) \ .$$

Theorem 2. Suppose that $E\mu(1)=1$, $E\mu^2(1)\ln\mu(1)<\infty$ and $m,n\to\infty$.

1) If $n-m\to\infty$, then $A(m,n,d) \sim (2d+1)m^2/(n-m)^3 B$;

2) If $(n-m)n^{-2/3}\to \infty$, then $Z_2(m,n,d) \overset{P}{\to} 0$;

3) If $n-m \sim Cn^{2/3}$, $C \in (0,\infty)$, then for $k\in N_0$

$$P\{Z_2(m,n,d)=k\,|\,\mu(m)>0\}\to \frac{1}{k!}\sqrt{\Delta}\int_0^\infty x^{k-1/2}e^{-x-\sqrt{x\Delta}}\,dx,$$

where $\Delta = BC^3/(2d+1)$.

The proof of this theorem uses results of Silverman and Brown (1978). For obtaining the limit theorem in the case $n-m=o(n^{2/3})$, we use results of Sevastyanov (1972) and Ambrosimov (1976).

ACKNOWLEDGEMENT

I am taking use of the opportunity to express my sincere thanks to Professor A. M. Zubkov for his helpful comments on my investigations during many years.

REFERENCES

Ambrosimov, A.S. (1976) Normal laws in the scheme of sums of dependent random variables considered by B. A. Sevastyanov. Theory Probab. Appl., 21, 183-188.

Asmussen, S., Hering, H. (1976) Strong limit theorems for supercritical migration-branching processes. Math. Scand., 39, 327-342.

Athreya, K. B., Parthasarathy, P. R., Sankaranarayanan, G. (1974) Supercritical age-dependent branching processes with immigration. J. Appl. Probab., 11, 695-702.

Badalbayev, I.,S., Zubkov, A., M. (1983) A limit theorem for the sequence of branching processes with immigration. Theory Probab. Appl., 28, 382-388.

Durham, S. D. (1971) A problem concerning generalized age-dependent branching processes with immigration. Ann. Math. Stat., 42, 1121-1123.

Eagleson, G.K. (1975) Martingale convergence to mixtures of infinitely divisible laws. Ann. Probab., 3, 557-562.

Foster, J. (1969) Branching processes involving immigration. Ph. D. Thesis, University of Wisconsin.

Foster, J., Williamson, J. (1971) Limit theorems for the Galton-Watson process with time dependent immigration. Z. Wahrschein. und Verw. Geb., 20, 227-235.

Nagayev, S. (1975) A limit theorem for branching processes with immigration. Theory Probab. Appl. 20, 178-180.

Nagayev, S., Asadullin, M. (1985) On certain scheme for summation of random number of independent random variables with applications to branching processes. Dokl. Akad. Nauk SSSR, 285 (2), 293-296.

Rahimov, I. (1986) Critical branching processes with infinite variance and decreasing immigration. Theory Probab. Appl., 31, 88-100.

Rahimov, I. (1987) A limit theorem for random sums of dependent indicators and its applications in the theory of branching processes Theory Probab.Appl., 32, 2, 290-298.

Seneta E. (1970) An explicit-limit theorem for the critical Galton-Watson process with immigration. JRSS 32, 149-152.

Sevastyanov, B. A. (1957) Limit theorems for a special type branching random processes. Theory Probab. Appl., 2, 339-348.

Sevastyanov, B.A. (1972) Poisson limit law in the scheme of sums of dependent random variables . Theory Probab. Appl., 17, 695-699.

Shurenkov, B. M. (1976) Two limit theorems for critical branching processes. Theory Probab. Appl., 21, 548-558.

Silverman, B. and Brown, T. (1978) Short distances, flat triangles and Poisson limits. J. Appl. Probab. 15, 815-825.

NON HOMOGENEOUS DECOMPOSABLE BRANCHING PROCESSES

I. Rahimov[1], F. Yildirim[2], A. Teshabaev[3]

Romanovski Math. Inst. Tashkent, and METU, Ankara, Turkey

ABSTRACT

In this paper, a non-homogeneous, decomposable and continuous-time Markov branching process with three types of particles has been considered. Assume that the first type of particles are immutable, particles of the second type may transmute into particles of the second and third types and particles of the third type only into those of the third type and probabilities of these transmutations are independent of time. This process can be considered as a two-type decomposable branching process with time-dependent immigration. Some limit theorems are proved for the number of particles, when reproduction processes are critical and intensities of the number of "immigrants" are decreasing.

Key words: Markov Branching Process, multi-type, decomposable, immigration, time-dependent.

We consider a non homogeneous branching process with particles of three types T_1, T_2 and T_3. Assume that within the time interval $(t, t+\Delta t)$, $\Delta t \to 0$, a particle of type T_i is transmuted into a collection of particles $w=(w_1, w_2, w_3)$ of types T_1, T_2, T_3 respectively with probabilities $\delta_i^w +$

Postal addresses:

1,3. Tashkent, 700142, Hodjaev Str., 29, Romanovski Math. Inst. of Uzbek AS.

2. Dept. of Statistics, 06531, METU, Ankara, Turkey.

$p_i^W(t)\Delta t + o(\Delta t)$, where $\delta_k^W = 1$ for $w_k=1$, $w_i=0$, $i \neq k$, and $\delta_k^W = 0$ otherwise. We shall assume that particles of the type T_1 are immutable, that is under any change a particle of type T_1 yields exactly one particle of type T_1 and a certain collection of particles of types T_2 and T_3 which cannot revert back into T_1 (Sevastyanov 1971).

We also assume that particles of type T_2 may transmute into particles of types T_2, T_3 and particles of type T_3 only into those of type T_3 while probabilities of these transmutations are independent of time. This means that our process has only transitions of the form $T_1 \rightarrow T_2 \rightarrow T_3$.

We denote by $\mu_{kj}^\tau(t)$ the number of particles of type T_j obtained from one particle of type T_k within the time interval (τ, t) and $\mu_{kj}(t) = \mu_{kj}^0(t)$. The process $\mu(t) = (\mu_{12}(t), \mu_{13}(t))$ can be interpreted as a two type decomposable branching process with time dependent immigration. In the paper of Chistyakov (1970) it was considered a time-homogeneous decomposable Markov branching process with particles of the final type. Later by Ogura (1975) and by Polin (1976, 1977) limit theorems are proved for the discrete and continuous time decomposable branching processes. The survey of Vatutin and Zubkov (1985) and papers of Vatutin and Sagitov (1988,1989) contain quite a complete list of references concerning decomposable processes. Some limit theorems for the process $\mu(t)$ will be proved in this paper, when the reproduction processes are critical and intensities of the number of immigrating particles are decreasing.

We introduce the generating functions

$$f_k(t,X) = \sum_W p_k^W(t) x_1^{w_1} x_2^{w_2} x_3^{w_3} \ , \quad X = (x_1, x_2, x_3) \ ,$$

with $f_k(t,X) = 0$ for $x_1 = x_2 = x_3 = 1$ and for any $t \in [0,\infty)$. In our assumptions these generating functions can be represented in the form

$$f_1(t,X)=x_1 g(t,x_2,x_3), \quad f_2(t,X)=f_2(x_2,x_3), \quad f_3(t,X)=f_3(x_3).$$

Introduce the generating functions

$$F(t,X) = \sum_W P\{\mu_{k1}^\tau(t)=w_1, \mu_{k2}^\tau(t)=w_2, \mu_{k3}^\tau(t)=w_3\} x_1^{w_1} x_2^{w_2} x_3^{w_3}$$

and put $H(t,x_2,x_3) = F_1^0(t,1,x_2,x_3)$. By the same arguments as Sevastyanov (1971) it can be shown that

$$H(t,x_2,x_3) = \exp\{ \int_0^t g(u, F_2(t-u,X), F_3(t-u,X))du \}. \tag{1}$$

We shall assume that, at the points $x_2=x_3=1$, the functions $f_2(X)$ and $f_3(X)$ have all derivatives up to the order three and let

$$\left. \frac{\partial f_k}{\partial x_j} \right|_{X=1} = a_{kj}, \quad \left. \frac{\partial^2 f_k}{\partial x_k^2} \right|_{X=1} = b_k, \quad k,j = 2,3.$$

We also put for $k,j = 2,3$

$$\left. \frac{\partial g(t,x_2,x_3)}{\partial x_k} \right|_{X=1} = \alpha_k(t), \quad \left. \frac{\partial^2 g(t,x_2,x_3)}{\partial x_k \partial x_j} \right|_{X=1} = \beta_{kj}(t),$$

and assume that

$$\sup_t \alpha_k(t) < \infty, \quad \sup_t \beta_{kj}(t) < \infty. \tag{2}$$

We consider a process in which $a_{22}=a_{33}=0$ and $\alpha_k(t) \to 0$ as $t \to \infty$, $k=2,3$. First we shall investigate the probability $P\{\mu(t) \neq 0\}$, $0=(0,0)$.

It is known from Savin and Chistyakov (1962), that

$$P\{\mu_{22}(t) + \mu_{23}(t) > 0\} \sim \frac{\sigma_2}{t^{1/2}}, \quad t \to \infty, \tag{3}$$

where $\sigma_2 > 0$. Using some arguments of Polin (1976) in our case, we obtain

$$\sigma_2 = 2[\frac{a_{23}}{b_2 b_3}]^{1/2} .$$

We assume that $\alpha_k(t)$, $k=2,3$ vary regularly as $t \to \infty$ and

$$\beta_{22}(t)\ln t \to 0 , \quad \beta_{ij}(t) \to 0 , \quad i+j \neq 4 . \tag{4}$$

Theorem 1. If $a_{22}=a_{33}=0$, $b_2>0$, $b_3>0$, condition (4) is satisfied and as $t \to \infty$

$$\alpha_2(t) \, t^{1/2} \overset{}{\to} C_2, \quad \alpha_3(t)\ln t \to C_3, \quad C_i \in [0,\infty) \tag{5}$$

then

$$P = \lim_{t \to \infty} P(\mu(t) \neq 0) = 1 - \exp\{ - 2\Pi C_2 \, [\frac{a_{23}}{b_2 b_3}]^{1/2} - \frac{2C_3}{b_3} \} .$$

It follows from this theorem that if $\max(C_2, C_3) = 0$, then $P=0$ and it can be shown that if $\max(C_2, C_3) = \infty$, then $P=1$.

Proof. Consider relation (1) with $x=x=0$. Expanding the function $g(u, F_2, F_3)$ in a neighborhood of the point $F_2=F_3=1$, we get

$$\ln P\{\mu(t)=0\} = -\sum_{i=2}^{3} \int_0^t \alpha_i(u)(1-F_i(t-u,0))du +$$

$$\frac{1}{2}\sum_{i,j=2}^{3} \int_0^t \hat{\beta}_{ij}(u,t)(1-F_i(t-u,0))(1-F_j(t-u,0))du, \tag{6}$$

where $0 \leq \hat{\beta}_{ij}(u,t) \leq \beta_{ij}(u)$ for any $t \in [0,\infty)$.

Let $C_i > 0$. Using (3) and (5), we obtain that as $t \to \infty$,

$$\int_0^{t-\ln t} \alpha_2(u)(1-F_2(t-u,0,0))du \sim 2C_2\sigma_2 \arc \sin\sqrt{1-\frac{\ln t}{t}} \, ,$$

$$\int_{t-\ln t}^{t} \alpha_2(u)(1-F_2(t-u,0,0))du \to 0 \, ,$$

and, consequently,

$$\lim_{t\to\infty} \int_0^t \alpha_2(u)(1-F_2(t-u,0,0))du = C_2\sigma_2\pi \, . \qquad (7)$$

Since $\mu_{33}(t)$ is a simple one dimensional critical branching process,

then

$$1-F_3(t,0) \sim \frac{2}{b_3 t} \, , \quad t\to\infty \, . \qquad (8)$$

Using (8) and the condition $\alpha_3(t) \sim C_3/\ln t$ it can be shown that the second

summand of the first sum in (6) has limit $-2C_3/b_3$.

It remains only to find the limit of the second sum in (6). To do

this, we use estimates $\hat{\beta}_{ij}(u,t) \leq \beta_{ij}(u)$. Using (3) in the summand with

$i=j=2$, we obtain that it is less than

$$\text{const.} \int_0^{t-t_0} \beta_{22}(u) \frac{du}{t-u}$$

for some $t_0 > 0$. But under the condition $\beta_{22}(t)\ln t = o(1)$, this integral is an

infinitesimal. For the other summands, if we use (3) and (8), we obtain

estimations of the form :

$$\text{const} \int_0^{t-t_0} \beta_{ij}(u) \frac{du}{(t-u)^{1+\alpha}} \, , \quad i+j\neq 4 \, ,$$

where $t_0 > 0$ and α is equal to either 1/2 or 1. If $\beta_{ij}(t)\to 0$, $t\to\infty$, $i+j\neq 4$, then

these integrals are also infinitesimal. Hence the limit of the second sum in

(6) is equal to zero .

The theorem is thus proved for $C_i > 0$. In the case when $C_i = 0$ it can be proved by using the similar arguments.

Now we consider the case

$$\alpha_2(t) \sim 1(t) / t^{1/2} , \quad t \to \infty \tag{9}$$

where $1(t) \to \infty$ and varies slowly as $t \to \infty$. In this case $P = \lim_{t \to \infty} P\{\mu(t) \neq 0\} = 1$ as was noted already .

Theorem 2. If $a_{22} = a_{33} = 0$, $a_{23} > 0$, $b_2 > 0$, $b_3 > 0$ and conditions (2), (4) and (9) are satisfied, then

$$\frac{2 \mu_{13}(t)}{b_3 a_{23} t (1 (t))^2} \overset{d}{\to} \xi$$

as $t \to \infty$, where ξ has a stable distribution with exponent $1/2$ and density function

$$P(x, b_2) = \frac{2}{bx(\pi x)^{1/2}} e^{-4/b_2^2 x} , \quad x > 0.$$

Proof. The density function $P(x, b_2)$ corresponds to the Laplace transform $\exp\{-4\sqrt{\lambda} / b_2\}$, (Feller 1968, Chapter 13). Hence it suffices to show that

$$H(t, 1, x_3(t)) \to \exp\{-4\sqrt{\lambda} / b_2\}$$

as $t \to \infty$, where $x_3(t) = \exp[-2\lambda/b_3 a_{23} t 1^2(t)]$, $\lambda > 0$.

Consider relation (1) with $x_2 = 1$, $x_3 = x_3(t)$. Expanding the function $g(u, F_2, F_3)$ in a neighborhood of the point $F_2 = F_3 = 1$, we get

$$\ln H(t, 1, x_3) = -\sum_{i=2}^{3} \int_0^t \alpha_i(u)(1 - F_i(t-u, X)) du + \varepsilon(t, \lambda) \tag{10}$$

where $X=(x_2,x_3)$ and $\varepsilon(t,\lambda)$ is less than the second sum in relation (6). It follows from the proof of Theorem 1 that, under our conditions, $\varepsilon(t,\lambda) \to 0$ as $t\to\infty$, $\lambda>0$.

It remains finding the limit of the first sum in (10). It is known (Chistyakov (1970)) that the function $R_2=1-F_2$ can be represented in the form

$$R_2(t,1,x_3) = Z_\gamma(t) - \frac{2\theta(t)}{2} [\sqrt{\frac{\gamma}{a_{23}}} - \int_0^t \theta(u)du]^{-1} , \qquad (11)$$

for all $t\geq 0$, $\gamma = 2/b_3(1-x_3)\geq\gamma_0>0$, where

$$\theta(t) = \exp\{-4\sqrt{a_{23}\gamma} \ (\sqrt{1 + \frac{t}{\gamma}} - 1) + O(\ln(1+ \frac{t}{\gamma}))\}$$

and $Z_\gamma(t)$ satisfies the inequality

$$|Z_\gamma(t) - \frac{2}{b_2} [\frac{a_{23}}{t+\gamma}]^{1/2} | \leq \frac{const}{t+\gamma} \qquad (12)$$

for all $t\geq 0$, $\gamma\geq\gamma_0>0$, $Z_\gamma(0)=2 [a_{23}/b_2\gamma]^{1/2}$.

First we shall consider the integral

$$I_1 = \int_0^t \frac{2\alpha_2(u)}{b_2} [\frac{a_{23}}{t+\gamma}]^{1/2}du = \int_0^t \varphi_t(u)du.$$

Since $l(t)$ is a slowly varying function, there is a function $\lambda_1(t)\to\infty$ as $t\to\infty$ such that for any function $\lambda(t)$, $1\leq\lambda(t)\leq\lambda_1(t)$,

$$\lim_{t\to\infty} l(t/\lambda(t))/l(t)=1. \qquad (13)$$

Then

$$\int_{t/\lambda_1}^{t} \varphi_t(u)du \sim \frac{(4a_{23})^{1/2}}{b_2} l(t) \int_{t/\lambda_1}^{t} \frac{du}{[u(t-u+\gamma)]^{1/2}}$$

as $t\to\infty$ and, since $\gamma \sim a_{23}tl^2(t)/\lambda$, $t\to\infty$, the last integral is equal to

$$2\arcsin[\frac{x}{t+\gamma}]^{1/2} \Big|_{t/\lambda}^{t} \sim 2 [\lambda/a_{23}l^2(t)]^{1/2}.$$

From the estimate

$$\int_0^{t/\lambda_1} \varphi_t(u)du \leq const.\int_0^{t/\lambda_1} \alpha_2(u)t^{-1/2} du,$$

we have that

$$\lim_{t\to\infty} I_1 = 4 \lambda^{1/2}/ b_2 .$$ (14)

It can also be shown by the similar arguments that

$$I_2 = \int_0^t \frac{\alpha_2(u)du}{t-u+\gamma} \to 0, \quad t\to\infty .$$ (15)

Now we shall consider the integral

$$I_3 = \int_0^t \alpha_2(u) \frac{2\theta(t-u)}{b_2} ((\frac{\gamma}{a_{23}})^{1/2} - \int_0^{t-u} \theta(x)dx)^{-1} du$$

We use the following estimates of the function $\theta(u)$, $0\leq u \leq t$, (Chistyakov (1970)) :

$$\theta(u)=\exp\{-2 (\frac{a_{23}}{\gamma})^{1/2} u \}[1+0(\frac{u}{\gamma} + \frac{u^2}{\gamma})],$$ (16)

$$\int_0^t \theta(u)du= \frac{1}{2}\sqrt{\frac{\gamma}{a_{23}}} (1-\exp\{-2 (\frac{a_{23}}{\gamma})^{1/2} t \}) +0(1)$$ (17)

as $t \to \infty$, $\gamma \sim t l^2(t) a_{23}/\lambda$. By the same arguments as in estimation of the integral I_1 using (16) and (17), we find that

$$\limsup_{t \to \infty} I_3 \leq \limsup_{t \to \infty} \{ \text{const.} \frac{l(t)}{t^{1/2}} \int_0^t \psi_t(u)du \}$$

where

$$\psi_t(u) = \frac{1}{u^{1/2}} \exp\{-2(\frac{a_{23}}{\gamma})^{1/2} (t-u)\} , \quad u > 0.$$

Let $L(t) \to \infty$, $L(t) = o(t)$, be a function such that $\alpha_2(t)L(t) \to \infty$. Using estimates

$$\int_0^{t-t/L(t)} \psi_t(u)du \leq \text{const.} \exp\{-2\sqrt{a_{23}} / \alpha_2(t)L(t)\}\sqrt{t} ,$$

and

$$\int_0^{t-t/L(t)} \psi_t(u)du \leq \text{const} \sqrt{t} / L(t) ,$$

we obtain that

$$\lim_{t \to \infty} I_3 = 0. \tag{18}$$

Since

$$|\int_0^t \alpha_2(u)R_2(t-u,1,x_3)du - I_1| \leq I_2 + I_3 ,$$

it follows from relations (14), (15) and (18) that the summand with $i=2$ of the first sum in (10) has the limit $-4\sqrt{\lambda} / b_2$.

The summand with $i=3$ in this sum is an infinitesimal because of the inequality

$$\int_0^t \alpha_3(u)R_3(t-u,x_3(t))du \leq \sup_u \alpha_3(u)t(1-x_3(t))$$

and the theorem is thus proved.

Limit theorems in the case when $l(t) = \text{const.}$ and $l(t) \to 0$ can be obtained by using the similar arguments.

REFERENCES

Chistyakov, V.P.(1970) Some limit theorems for branching processes with particles of final type. Theory Probab.Appl., 15, 515-521.

Feller, W. (1968) An Introduction to Probability Theory and its Applications Vol.2, Wiley, New York.

Ogura, Y. (1975) Asymptotic behavior of multitype Galton-Watson processes. J. Math. Kyoto Univ., 15, 251-302.

Polin, A. K. (1976) Limit theorems for decomposable critical branching processes. Matematicheski Sbornik, 100, 420-435.

Polin, A. K. Limit theorems for decomposable branching processes with particles of final type. Matematicheski Sbornik, 104, 151-161.

Savin, A.A. and Chistyakov,V.P. (1962) Some theorems for branching processes with several types of particles.Theory Probab.Appl., 7, 93-100.

Sevastyanov, B. A. (1971) Branching Processes, Nauka, Moscow.

Vatutin, V. A. and Zubkov, A. M. (1985) Branching Processes,1, In :Progress in Science and Technology: Probability Theory .Math. Statist. Theor. Cybernetics, 23 , 3-67 .

Vatutin, V.A. and Sagitov, S.M. (1988, 1989) Decomposable critical branching Bellman-Harris process with particles of two different types 1, 2. Theory Probab. Appl., 33, 495-507, 34, 251-262.

AGE-DEPENDENT BRANCHING PROCESSES WITH STATE-DEPENDENT IMMIGRATION

MAROUSSIA N. SLAVTCHOVA-BOJKOVA   AND   NICKOLAY M. YANEV [1]

Abstract

We consider a model of Bellman-Harris branching processes with immigration only in the state zero (BHIO) which permits in addition an immigration component of i.i.d. BHIO processes entering the population at i.i.d. times of an independent renewal process (BHIOR). Asymptotic properties and limit theorems are proved in non-critical cases.

age-dependent branching processes; state-dependent immigration; limit theorems
60J80

1   Introduction.

Branching processes with immigration were first introduced and studied by Sevastyanov (1957) in a continuous-time Markov case. Later Jagers (1968) considered Bellman - Harris branching processes with a renewal immigration component and Yanev (1972) obtained limit theorems in a class of decomposable age-dependent branching processes with immigration.

A model with a state-dependent immigration component was first investigated by Foster (1971) and Pakes (1971, 1975). They considered a modification of the Bienaymé - Galton - Watson process allowing immigration whenever the number of particles is zero. The continuous-time analogue of this process was studied by Yamazato (1975).

BHIO (Bellman-Harris branching processes with immigration only in the state zero) were introduced and investigated in the critical case by Mitov and Yanev (1985, 1989). Their asymptotic results generalised those obtained by Foster (1971) and Yamazato (1975) for the critical Markov processes.

---

[1] Postal address for both authors: Department of Probability and Statistics, Institute of Mathematics, Bulgarian Academy of Sciences, 1113 Sofia, Bulgaria.
Supported by the National Foundation for Scientific Investigations, grant MM-4/91.

BHIOR processes (i.e. BHIO processes which admit in addition a renewal immigration component) were considered by Weiner (1991). He proved that in the critical case the rate of convergence in probability is $Ct^2/\log t$ as $t \to \infty$ (C is a certain constant). In contrast to this, we obtain that the subcritical processes have a linear growth (Theorem 5.1).

K. Athreya (1969) first showed an analog of the classical Kesten and Stigum theorem (see Athreya and Ney (1972)) for supercritical Bellman-Harris branching processes which refine and make more precise the estimates of the growth of the processes on the set of non-extinction. Our aim is to generalize Athreya's result (1969) for the BHIOR model, and it is shown that for these processes the asymptotic results have an analogy with those obtained by Athreya.

On the other hand, Kaplan and Pakes (1974) studied the supercritical Bellman-Harris processes, where the immigrants enter the population at the event times of an ergodic renewal process. The techniques they used to prove convergence in distribution and almost sure (a.s.) convergence, depend only on the the asymptotics of the underlying age-dependent processes and we apply their approach to study the more general model with two types of state-dependent immigration.

In the present paper we establish mean square convergence of the supercritical BHIOR processes to a random variable, characterized by an integral equation of its Laplace transform (L.T.) via direct renewal methods (Theorem 5.2). On the other hand, under the classical $X \log X$ condition a convergence in distribution is proved in Theorem 5.3 and the properties of the limit L.T. are given in Theorem 5.4. Almost sure convergence is also obtained under the second-moment conditions (Theorem 5.5).

Finally, we would like to mention that the limit results for the multitype case will be presented in a forthcoming paper.

2 Model.

Let $\{X(t)\}_{t \geq 0}$ be a population process, wherein the individuals reproduce according to a BHIO process and in addition a random number of immigrants enter the population at the event times τ_i of a given renewal process. It is assumed that the interarrival times $T_1 = \tau_1, T_2 = \tau_2 - \tau_1, \ldots$ are i.i.d.r.v. with common distribution (d.f.) $G_0(t)$. We will always assume that $\int_0^\infty t\, dG_0(t) < \infty$. The τ_i and the numbers ν_i of immigrants arriving at each immigration are assumed to be independent. The ν_i are i.i.d. with common p.g.f. $f_0(s)$. The BHIO process $\{Z(t)\}_{t \geq 0}$ is governed by a life-time distribution $G(t)$, an offspring p.g.f. $h(s)$, a p.g.f. $f(s)$ of the random number Y_i of immigrants in the state zero and the d.f. $K(t)$ of the duration X_i of staying in the state zero. It is assumed that $\int_0^\infty t\, dK(t) < \infty$.

We will use the definition given by Mitov and Yanev (1985):

$$Z(t) = Z_{N(t)+1}(t - S_{N(t)} - X_{N(t)+1})\mathbb{I}_{\{S_{N(t)}+X_{N(t)+1} < t\}}, \quad Z(0) = 0,$$

where $\{Z_i(t)\}$ are independent Bellman-Harris branching processes starting with a random number Y_i of particles, $N(t) = \max\{n \geq 0 : S_n \leq t\}$, $S_0 = 0$, $S_n = \sum_{i=1}^{n} U_i$, $U_i = X_i + \sigma_i$, $\sigma_i = \inf\{t : Z_i(t) = 0\}$.

We introduce the following notation for the p.g.f. of the local characteristics of the process:

$$f(\dot{s}) = \mathbb{E}s^{Y_1} = \sum_{k=1}^{\infty} f_k s^k, \quad h(s) = \sum_{k=0}^{\infty} p_k s^k, \quad f_0(s) = \mathbb{E}s^{\nu_1} = \sum_{k=0}^{\infty} q_k s^k.$$

It will be assumed that:

$$0 < A = h'(1) < \infty, \quad m = f'(1) < \infty, \quad m_0 = f_0'(1) < \infty, \tag{2.1}$$

$$G(t), \ G_0(t) \ \text{and} \ K(t) \ \text{are non-lattice}, \tag{2.2}$$

$$0 < B = h''(1) < \infty, \quad n = f''(1) < \infty, \quad f_0''(1) = b_2 < \infty, \tag{2.3}$$

$$r = \int_0^{\infty} x \, dG(x) < \infty, \quad a = \int_0^{\infty} x \, dK(x) < \infty, \quad r_0 = \int_0^{\infty} x \, dG_0(x) < \infty. \tag{2.4}$$

Note that $L(t) = \mathbb{P}\{X_i + \sigma_i \leq t\}$ is non-lattice with $L(0) = 0$.

3 Equations.

Let $\Phi(t, s) = \mathbb{E}s^{Z(t)}$, $\Phi(0, s) = 1$, $\Phi_0(t, s) = \mathbb{E}s^{X(t)}, \Phi_0(0, s) = 1$, $\Phi_0(t, \tau, s_1, s_2) = \mathbb{E}\{s_1^{X(t)} s_2^{X(t+\tau)}\}$.

We will use the notation $\hat{Z}(t)$ for an ordinary Bellman-Harris branching process with p.g.f. $F(t, s) = \mathbb{E}s^{\hat{Z}(t)}$, $F(0, s) = s$.

It is known, that the p.g.f. $\Phi_0(t, s)$ admits the representation (see Weiner (1991))

$$\Phi_0(t, s) = \int_0^t f_0(\Phi(t - u, s))\Phi_0(t - u, s) \, dG_0(u) + 1 - G_0(t), \tag{3.1}$$

where the p.g.f. $\Phi(t, s)$, satisfies the renewal equation (see Mitov and Yanev (1985))

$$\Phi(t, s) = \int_0^t \Phi(t - u, s) \, dL(u) + 1 - K(t) - L(t) + \int_0^t f(F(t - u, s)) \, dK(u), \tag{3.2}$$

and $F(t, s)$ is the unique solution of the equation

$$F(t, s) = \int_0^t h(F(t - u, s)) \, dG(u) + s(1 - G(t)), \quad F(0, s) = s. \tag{3.3}$$

By the law of total probabilities it is not difficult to obtain the equation

$$\begin{aligned}
\Phi_0(t, \tau, s_1, s_2) = \ & \int_0^t \Phi_0(t - u, \tau, s_1, s_2) f_0(\Phi(t - u, \tau, s_1, s_2)) \, dG_0(u) \\
& + \int_0^{t+\tau} \Phi_0(t + \tau - u, s_2) f_0(\Phi(t + \tau - u, s_2)) \, dG_0(u) \\
& + (1 - G_0(t + \tau)),
\end{aligned} \tag{3.4}$$

where $\Phi(t, \tau, s_1, s_2) = \mathbb{E}s_1^{Z(t)} s_2^{Z(t+\tau)}$, $|s_i| \leq 1$, $i = 1, 2$.

4 Moments.

Denote the moments $M_0(t) = \mathbb{E}X(t)$, $M_{02}(t) = \mathbb{E}X(t)(X(t)-1)$, $M(t) = \mathbb{E}Z(t)$, $M_2(t) = \mathbb{E}Z(t)(Z(t)-1)$, $N(t,\tau) = \mathbb{E}X(t)X(t+\tau)$.

As usual the Malthusian parameter α is the unique root of the equation $A \int_0^\infty e^{-\alpha u} dG(u) = 1$.

Theorem 4.1 *Under the assumptions* (2.1), (2.2) *and* (2.4) *if $A < 1$,* $0 < h^{(n)}(1) < \infty$, $0 < f^{(n)}(1) < \infty$, *and* $\int_0^\infty e^{-\alpha y} dG(y) < \infty$, *then* $\mathbb{E}[Z^n(t)] \to c_n$, $c_n < \infty$ *as $t \to \infty$.*

Proof. For the Bellman-Harris process, from (3.3) and using the Faà di Bruno's formula (see Abramowitz and Stegun (1970), p.823), it follows that $\mathbb{E}[\hat{Z}^n(t)] \equiv \hat{M}_n(t)$ satisfies the equation
$\hat{M}_n(t) = \int_0^t \sum_{m=0}^n h^{(m)}(1) \sum \frac{n!}{(1!)^{a_1} a_1! \ldots (n!)^{a_n} a_n!} [\hat{M}_1(t-u)]^{a_1} \ldots [\hat{M}_n(t-u)]^{a_n} dG(u)$, where the summation is over

$$a_1 + 2a_2 + \ldots + na_n = n \quad \text{and} \quad \sum_{i=1}^n a_i = m. \tag{4.1}$$

Hence using induction by $n \geq 1$ and the method of Theorem 4.2 (see Slavtchova and Yanev (1991)) we can conclude that

$$\hat{M}_n(t) \sim \hat{c}_n e^{\alpha n t}, \quad n \geq 1, \quad \hat{c}_n < \infty. \tag{4.2}$$

Differentiating (3.2) one obtains

$$\left. \frac{\partial^n \Phi(t,s)}{\partial s^n} \right|_{s=1} = \int_0^t \left. \frac{\partial^n \Phi(t-u,s)}{\partial s^n} \right|_{s=1} dL(u) + \int_0^t \left. \frac{\partial^n f(F(t-u,s))}{\partial s^n} \right|_{s=1} dK(u).$$

Therefore again applying Faà di Bruno's formula for the moment $\mathbb{E}[Z^n(t)] \equiv M_n(t)$ we get

$$
\begin{aligned}
M_n(t) &= \int_0^t M_n(t-u) dL(u) \\
&\quad + \int_0^t \sum_{m=0}^n f^{(m)}(1) \sum \frac{n! [\hat{M}_1(t-u)]^{a_1} \ldots [\hat{M}_n(t-u)]^{a_n}}{(1!)^{a_1}(a_1!) \ldots (n!)^{a_n}(a_n)!} dK(u) \\
&= \int_0^t M_n(t-u) dL(u) + R(t),
\end{aligned}
$$

where the summation is over (4.1). Hence we obtain the asymptotic (see (4.2))

$$M_n(t) \sim \frac{\int_0^\infty R(t) dt}{\int_0^\infty [1 - L(t)] dt} = c_n < \infty,$$

and the proof is completed.

Theorem 4.2 *Under the assumptions of Theorem 4.1, if in addition $f_0(s)$ is analytic,* $|s| < \delta + 1$, $\delta > 0$, *then*

$$\mathbb{E}X^n(t) \sim \left\{ \frac{m m_0 r t}{\nu_0 (1-A) r_0} \right\}^n, \tag{4.3}$$

where $\nu_0 = \int_0^\infty t dL(t) < \infty$.

Proof. A Taylor expansion yields

$$\Phi_0(t,s) = 1 - G_0(t) + \int_0^t (\Phi_0(t-u,s) - 1)dG_0(u)$$

$$+ \sum_{n=1}^\infty \int_0^t \frac{(-1)^n f_0^{(n)}(1)}{n!}(1 - \Phi(t-u,s))^n \Phi_0(t-u,s)dG_0(u).$$

Taking Laplace transforms and re-inverting, it follows that

$$\Phi_0(t,s) = 1 + m_0 \int_0^t (\Phi(t-u,s) - 1)\Phi_0(t-u,s)dH_0(u)$$

$$+ \sum_{l=1}^\infty \int_0^t \frac{(-1)^l f_0^{(l)}(1)}{l!}(1 - \Phi(t-u,s))^l \Phi_0(t-u,s)dH_0(u), \tag{4.4}$$

where $H_0(t) = \sum_{l=0}^\infty G_0^{*l}(t)$.

As $M(t) = \mathbb{E}Z(t)$ one obtains $M_0(t) = m_0 \int_0^\infty M(t-u)dH_0(u)$ and therefore

$$M_0(t) \sim m_0 \int_0^t \frac{mr}{(1-A)\nu_0}du = \frac{mm_0r}{(1-A)r_0\nu_0}t, \quad t \to \infty.$$

Let $M_{0n}(t) = \mathbb{E}\{X^n(t)\}$. Then considering orders of magnitude of t, $(t \to \infty)$, it follows by induction that for $n \geq 2$,

$$M_{0n}(t) \sim nm_0 \int_0^t M(t-u)M_{0n-1}(t-u)dH_0(u).$$

Again using induction, assume the result of the theorem for $n-1$. Then by the methods of Mitov and Yanev(1985, Theorem 2, p.761) one obtains for $n \geq 2$,

$$M_{0n}(t) \sim nm_0 \int_0^t \frac{mr}{(1-A)\nu_0 r_0}\left[\frac{mm_0rt}{(1-A)r_0\nu_0}\right]^{n-1}du = \left\{\frac{mm_0rt}{(1-A)r_0\nu_0}\right\}^n,$$

completing the induction and establishing the result.

It is well-known (see Harris (1963), Ch. VI, Theorem 17.1), that for the supercritical Bellman-Harris processes

$$\lim_{t \to \infty} \mathbb{E}\hat{Z}(t)e^{-\alpha t} = A_0 = \frac{\int_0^\infty e^{-\alpha u}(1 - G(u))du}{A \int_0^\infty u e^{-\alpha u}dG(u)}. \tag{4.5}$$

Denote

$$\bar{A} = \frac{mA_0 \int_0^\infty e^{-\alpha u}dK(u)}{1 - L \int_0^\infty e^{-\alpha u}d\tilde{L}(u)} \tag{4.6}$$

where $\tilde{L}(t) = L(t)/L$, $L = L(+\infty)$.

Theorem 4.3 *Under the assumptions (2.1) and (2.2) when $A > 1$,*

$$\lim_{t \to \infty} M_0(t)e^{-\alpha t} = \bar{A}_0 = \frac{m_0\bar{A}\Delta(\alpha)}{1 - \Delta(\alpha)}, \quad \Delta(\lambda) = \mathbb{E}e^{-\lambda T_1}. \tag{4.7}$$

Theorem 4.4 *Let the conditions (2.1) – (2.3) hold and $A > 1$. Then*

$$\lim_{t\to\infty} M_{02}(t)e^{-2\alpha t} = \bar{N}_0 = \frac{(m_0\bar{N} + b_2\bar{A}^2 + 2m_0\bar{A}\bar{A}_0)\int_0^\infty e^{-2\alpha u}dG_0(u)}{1 - \int_0^\infty e^{-2\alpha u}dG_0(u)}, \qquad (4.8)$$

where $B_0 = \dfrac{BA_0\int_0^\infty e^{-2\alpha u}dG(u)}{1 - A\int_0^\infty e^{-2\alpha u}dG(u)}$, $\bar{N} = \dfrac{(mB_0 + nA_0^2)\int_0^\infty e^{-2\alpha u}dK(u)}{1 - L_0\mu_0}$,

$\mu_0 = \displaystyle\int_0^\infty e^{-2\alpha u}d\tilde{L}(u) < 1$ *and* A_0, $\bar{A}$, $\bar{A}_0$ *are given by (4.6), (4.7) and (4.8).*

Theorem 4.5 *Assume the conditions of Theorem 4.4. Then* $N_0(t,\tau) = e^{\alpha(2t+\tau)}\bar{N}_0(1 + o(1))$, *uniformly for* $\tau \geq 0$, *where* $\bar{N}_0$ *is defined by (4.9).*

The proofs of Theorems 4.3, 4.4 and 4.5 follow from a similar renewal approach to Theorem 3.2 (see Slavtchova and Yanev (1991)). The detailed proofs will appear in Slavtchova-Bojkova and Yanev (1994).

5 Limit theorems.

Theorem 5.1 *Under the assumptions of Theorem 4.2* $X(t)/t \xrightarrow{\mathbb{P}} mm_0r/(1 - A)\nu_0r_0$ *as* $t \to \infty$.

Proof. From (4.3) we get $\mathbb{E}\{X(t)/t^n\} \sim mm_0r/(1 - A)\nu_0r_0$, $n \geq 1$. Applying the method of the moments one obtains that $X(t)/t$ converges in probability to a constant r.v., whose distribution is determined by the asymptotic moments of the process.

Note that in the critical case (A=1) Weiner (1991) was proved that $X(t)\log t/t^2$ converges in probability to a certain constant.

Theorem 5.2 *Assume the conditions of Theorem 4.4. If* $\alpha > 0$ *is the Malthusian parameter, then the process* $W(t) = X(t)/\mathbb{E}X(t)$, *converges in mean square to a r.v.* W *whose L.T.* $\varphi(\lambda) = \mathbb{E}e^{-\lambda W}$, $\lambda \geq 0$ *satisfies the equation:*

$$\varphi(\lambda) = \int_0^\infty \varphi(\lambda e^{-\alpha u})f_0(\psi(\lambda e^{-\alpha u}))dG_0(u), \qquad (5.1)$$

and $\psi(\lambda)$ *is the unique solution of the equation*

$$\psi(\lambda) = \int_0^\infty \psi(\lambda e^{-\alpha u})dL(u) + \int_0^\infty f(\theta(\lambda e^{-\alpha u}))dK(u) - f(q) \qquad (5.2)$$

where $q = \displaystyle\lim_{t\to\infty} \mathbb{P}\{\hat{Z}(t) = 0|\hat{Z}(0) = 1\}$ *and* $\theta(\lambda)$ *is the unique solution of the equation*

$$\theta(\lambda) = \int_0^\infty h(\theta(\lambda e^{-\alpha u}))dG(u), \qquad (5.3)$$

in the class

$$C = \left\{\theta : \theta(u) = \int_0^\infty e^{-ut}dF(t),\, F(0^+) < 1,\, \int_0^\infty t\,dF(t) = 1\right\}.$$

Proof. Using Theorems 4.3, 4.4 and 4.5 it is not difficult to show that $\lim_{t \to \infty} \{W(t + \tau) - W(t)\}^2 = 0$ uniformly for $\tau > 0$, which is equivalent to the mean square convergence to a r.v. W. The rest of the argument directly follows by the results of Slavtchova and Yanev (1991).

Corollary 5.1 *Under the conditions of Theorem 4.4,* $EW = \bar{A}_0$, $VarW = \bar{N}_0 - \bar{A}_0^2$, *where $\bar{A}_0$ and $\bar{N}_0$ are constants, defined by (4.8) and (4.9).*

Proof. The result follows immediately by differentiating (5.1) and setting $\lambda = 0$.

Our next theorem aims to provide an analogue to the result of Athreya (1969) for the Bellman-Harris branching processes.

Theorem 5.3 *Assume either $m_0 < \infty$ or $\sum_{j=2}^{\infty} j p_j \log j < \infty$. Then*

$$W(t) = X(t)/\bar{A}_0 e^{\alpha t} \xrightarrow{d} W, \ \bar{A}_0 \text{ is defined by (4.7) and the r.v. } W \text{ has a L.T.}$$

$$\varphi(\lambda) = \mathbb{E}\{e^{-\lambda W}\} = \mathbb{E}\left\{\prod_{i=1}^{\infty} f_0(\psi(\lambda e^{-\alpha \tau_i}))\right\}, \tag{5.4}$$

satisfying the equation (5.1), where $\psi(\lambda)$ is defined by (5.2).

The properties of $\varphi(\lambda)$ are given with the next theorem.

Theorem 5.4 *(I) Assume $\sum_{j=2}^{\infty} j p_j \log j < \infty$: (a) If $\sum_{j=2}^{\infty} q_j \log j = \infty$, then*

$\varphi(\lambda) \equiv 0$; *(b) If $\sum_{j=2}^{\infty} q_j \log j < \infty$, then $\varphi(\lambda)$ is the L.T. of r.v. which has a continuous distribution on $[0, \infty)$.*

(II) Assume $m_0 < \infty$: (c) If $\sum_{j=2}^{\infty} j p_j \log j = \infty$, then $\varphi(\lambda) \equiv 1$; (d) If $\sum_{j=2}^{\infty} j p_j \log j < \infty$, then $\varphi(\lambda)$ is the L.T. of a r.v., such that: (i) $\mathbb{E}\{W\} = \bar{A}_0$; (ii) $\mathbb{P}\{W = 0\} = 0$; (iii) W has an absolutely continuous distribution on $[0, \infty)$ with a continuous density function.

Theorem 5.5 *Assume that $\sum_{j=1}^{\infty} j^2 p_j < \infty$ and $\sum_{j=1}^{\infty} j^2 q_j < \infty$. Then $W(t) \xrightarrow{a.s.} W$.*

The proofs of the last three theorems use the following representation of the process $\{X(t)\}$. Let $\{Z_{li}(t)\}_{t \geq 0, l, i \geq 1}$ be a set of i.i.d. stochastic processes defined on a common probability space, each having the same distribution as the BHIO process $\{Z(t)\}_{t \geq 0}$. Furthermore, all of these processes are assumed independent of the event times $\{\tau_i\}$ and the number of immigrants $\{\nu_i\}$. Define the renewal process $n(t) = \max\{n : \tau_n \leq t\}$. Then

$$X(t) = \sum_{i=1}^{n(t)} \sum_{l=1}^{\nu_i} Z_{il}(t - \tau_i). \tag{5.5}$$

6   Proof of Theorem 5.3.

¿From (5.5) it follows that $\varphi(\lambda, t) = \mathbb{E}\{e^{-\lambda X(t)}\} = \mathbb{E}\left\{\prod_{i=1}^{n(t)} f_0(\psi(\lambda, t - \tau_i))\right\}$, where $\psi(\lambda, t) = $

$\mathbb{E}\{e^{-\lambda Z(t)} | Z(0) = 0\}$.

To prove Theorem 5.3 it is sufficient to show that, for any $\varepsilon > 0$, there exists an integer $N(\varepsilon)$ such that

$$\limsup_{t\to\infty} \left| \varphi(\lambda/\bar{A}e^{-\alpha t}, t) - \mathbb{E}\left\{\prod_{i=1}^{N(\varepsilon)} f_0(\psi(\lambda e^{-\alpha \tau_i}))\right\} \right| < \varepsilon. \qquad (6.1)$$

On the other hand, to prove (6.1) it is enough to show that, for any $\varepsilon > 0$, there exist numbers $N(\varepsilon)$ and $t(\varepsilon)$ and function $g(\varepsilon)$ such that the set

$$A(\varepsilon) = \left\{ \omega : n(t(\varepsilon)) > N(\varepsilon), \inf_{t > t(\varepsilon)} \left[\prod_{i=N(\varepsilon)+1}^{n(t)} f_0(\psi(\lambda/\bar{A}e^{-\alpha t}, t - \tau_i)) \right] > g(\varepsilon) \right\}$$

has probability greater than $1 - \varepsilon$, where the function $g(\varepsilon)$ satisfies: $0 \leq g(\varepsilon) \leq 1$ for ε sufficiently small and $\lim_{\varepsilon \to 0} g(\varepsilon) = 1$.

Using the idea of Kaplan and Pakes (1974) (see Lemma 2.1) with $\lim_{t\to\infty} \psi(\lambda/\bar{A}e^{\alpha t}, t) = \psi(\lambda), \lambda \geq 0$ (see Slavtchova and Yanev (1991)) then we have $\lim_{t\to\infty} f_0(\psi(\lambda/\bar{A}e^{\alpha t}, t - \tau_i)) = f_0(\psi(\lambda e^{-\alpha \tau_i}))$ a.s. for each i. Hence

$$\mathbb{E}\left\{ \mathbb{I}_{A(\varepsilon)} \prod_{i=1}^{N(\varepsilon)} f_0(\psi(\lambda e^{-\alpha \tau_i})) \right\} \geq g(\varepsilon)\mathbb{E}\left\{ \mathbb{I}_{A(\varepsilon)} \prod_{i=1}^{N(\varepsilon)} f_0(\psi(\lambda e^{-\alpha \tau_i})) \right\}.$$

Since $\mathbb{P}[A(\varepsilon)] > 1 - \varepsilon$, $\lim_{\varepsilon \to 0} g(\varepsilon) = 1$ and ε is arbitrary, then (6.1) follows.

Assume $m_0 < \infty$. From the Egorov's theorem for every $\varepsilon > 0$ there exists an $N(\varepsilon)$ such that

$$\mathbb{P}\left\{ \omega : \sum_{i=N(\varepsilon)}^{\infty} e^{-\alpha \tau_i} \leq \varepsilon^2 \right\} > 1 - \varepsilon.$$

If we define

$$A(t) = \left\{ \omega : \sum_{i=N(\varepsilon)}^{\infty} e^{-\alpha \tau_i} \leq \varepsilon^2, n(t) > N(\varepsilon) \right\}, \quad t > 0, A = \left\{ \omega : \sum_{i=N(\varepsilon)}^{\infty} e^{-\alpha \tau_i} < \varepsilon^2 \right\},$$

since $n(t) \to \infty$ a.s., then $\lim_{t\to\infty} \mathbb{P}[A(t)] = \mathbb{P}(A) > 1 - \varepsilon$. We can therefore choose $t(\varepsilon)$ such that $\mathbb{P}[A(t(\varepsilon))] > 1 - \varepsilon$. Suppose $\omega \in A(t(\varepsilon))$ and let ξ be a r.v., whose L.T. is $\prod_{i=N(\varepsilon)}^{n(t)} f_0(\psi(\lambda, t - \tau_i))$, $t > t(\varepsilon)$. It is clear (see Slavtchova and Yanev (1991)), that there exists a positive constant k such that $-\frac{d\psi(t,0)}{du} \leq ke^{\alpha t}$, where $\psi(u, t) = \mathbb{E}e^{-uZ(t)}$. Using that $\mathbb{E}\xi \leq e^{\alpha t}km_0 \sum_{i=N(\varepsilon)}^{\infty} e^{-\alpha \tau_i}$ and applying Markov's inequality we obtain that

$$\mathbb{P}\{\xi/\bar{A}e^{\alpha t} > \varepsilon m_0 k/\bar{A}|A(t(\varepsilon))\} \leq (1/\varepsilon) \sum_{i=N(\varepsilon)}^{\infty} e^{-\alpha \tau_i}$$

It follows then that $\mathbb{E}\{e^{-\lambda\xi/\bar{A}e^{\alpha t}}\} \geq e^{-\varepsilon k_1}[1 - \varepsilon]$, where $k_1 = \lambda m_0 k/\bar{A}$. If we choose $g(\varepsilon) = e^{-\varepsilon k_1}[1 - \varepsilon]$ then we are done.

Let us now assume that $\sum_{j=2}^{\infty} jp_j \log j < \infty$. Then

$$\limsup_{t\to\infty} \varphi(\lambda/\bar{A}e^{\alpha t}, t) \leq \mathbb{E}\left\{\prod_{i=1}^{\infty} f_0(\psi(\lambda/\bar{A}e^{-\alpha\tau_i}))\right\}. \tag{6.2}$$

When $\sum_{j=2}^{\infty} q_j \log j = \infty$, we have $\varphi(\lambda) \equiv 0$. Due to (6.2) we can assume $\sum_{j=2}^{\infty} q_j \log j < \infty$ without loss of generality. Choose $0 < \varepsilon < \frac{1}{2}\lambda_0$. From Egorov's theorem there exists a constant $N(\varepsilon)$ such that $\mathbb{P}\{\omega : \tau_i > i(\lambda_0 - \varepsilon), i \geq N(\varepsilon) \} > 1 - \varepsilon$, where $\lambda_0 = \mathbb{E}T_1$. Pick $t(\varepsilon)$ so large that

$$\mathbb{P}\{\omega : \tau_i > i(\lambda_0 - \varepsilon), \quad i \geq N(\varepsilon), \quad n(t(\varepsilon)) > N(\varepsilon)\} > 1 - \varepsilon. \tag{6.3}$$

Using the Mean Value Theorem , it is not difficult to show that there exists a constant D such that $\psi(\lambda/\bar{A}e^{\alpha t}, t - \tau_i) \geq 1 - -D\lambda e^{-\alpha\tau_i}, i \geq 1$. Let $t > T(\varepsilon)$. Then on the set defined by (6.3),

$$\prod_{i=N(\varepsilon)}^{n(t)} f_0(\psi(\lambda/\bar{A}e^{\alpha t}, t - \tau_i)) \geq \prod_{i=N(\varepsilon)}^{n(t)} f_0(1 - D\lambda e^{-i\alpha\lambda_0/2}).$$

We choose $N(\varepsilon)$ so large that $1 - D\lambda e^{-i\alpha\lambda_0/2} > 0, \quad i \geq N(\varepsilon)$. Define $g(\varepsilon) = \prod_{i=N(\varepsilon)}^{\infty} f_0(1 - D\lambda e^{-i\alpha\lambda_0/2})$. As the infinite product on the right converges, then $\lim_{\varepsilon\to 0} g(\varepsilon) = 1$. Since $N(\varepsilon) \to \infty$ as $\varepsilon \to 0$, we are done.

7 Proof of Theorem 5.4.

Assume $\sum_{j=2}^{\infty} jp_j \log j < \infty$. Under this assumption Slavtchova and Yanev (1991) proved that $\psi(\lambda)$ is the L.T. of a r.v. with mean 1. It follows from the Mean Value Theorem that there exists a $0 < \delta < 1$, independent of i, such that a.s. $1 - \lambda e^{-\alpha\tau_i} \leq \psi(\lambda e^{-\alpha\tau_i}) \leq 1 - \delta e^{-\alpha\tau_i}, i \geq 1$. Let $0 < \varepsilon < \lambda_0 = \mathbb{E}T_1$. Then there exists an integer $I(\varepsilon)$, depending only on the sample path, such that $i(\lambda_0 - \varepsilon) \leq \tau_i \leq i(\lambda_0 + \varepsilon), i \geq I(\varepsilon)$. Hence

$$1 - \lambda e^{-i\alpha(\lambda_0-\varepsilon)} \leq \psi(\lambda e^{-\alpha\tau_i}) \leq 1 - \delta e^{-i\alpha(\lambda_0+\varepsilon)}, \quad i \geq I(\varepsilon). \tag{7.1}$$

Thus

$$\sum_{i=I(\varepsilon)}^{\infty} \left[1 - f_0(\psi(\lambda e^{-\alpha\tau_i}))\right] \geq \sum_{j=1}^{\infty} q_j \left[\sum_{i=I(\varepsilon)}^{\infty} [1 - (1 - \delta e^{-i\alpha(\lambda_0+\varepsilon)})^j]\right].$$

Similarly

$$\sum_{i=I(\varepsilon)}^{\infty} \left[1 - f_0(\psi(\lambda e^{-\alpha\tau_i}))\right] \leq \sum_{j=1}^{\infty} q_j \left[\sum_{i=I(\varepsilon)}^{\infty} [1 - (1 - \lambda e^{-i\alpha(\lambda_0-\varepsilon)})^j]\right].$$

It is not difficult to show that $\sum_{i=I(\varepsilon)}^{\infty} [1-(1-\lambda e^{-i\alpha(\lambda_0-\varepsilon)})^j]$ and $\sum_{i=I(\varepsilon)}^{\infty} [1-(1-\lambda e^{-i\alpha(\lambda_0+\varepsilon)})^j]$ are asymptotic to $\log j$. Thus

$$\sum_{i=1}^{\infty} \left[1 - f_0(\psi(\lambda e^{-\alpha\tau_i}))\right] = \infty \ a.s. iff \sum_{j=2}^{\infty} q_j \log j = \infty.$$

Also note that $\varphi(\lambda) = 0 \Longleftrightarrow \sum_{i=1}^{\infty} \left[1 - -f_0(\psi(\lambda e^{-\alpha\tau_i}))\right] = \infty \ a.s.$ Theorem 5.4 (Ia) now follows immediately.

Using (7.1) we obtain

$$\prod_{i=I(\varepsilon)}^{\infty} f_0(\psi(\lambda e^{-\alpha\tau_i})) \geq \exp\left[\sum_{i=I(\varepsilon)}^{\infty} \log f_0(1 - \lambda e^{-i\alpha(\lambda_0-\varepsilon)})\right].$$

We can choose $I(\varepsilon)$ so large that $1 - f_0(1 - \lambda e^{-i\alpha(\lambda_0-\varepsilon)}) < \frac{1}{2}, \quad i \geq I(\varepsilon)$. Thus

$$\sum_{i=I(\varepsilon)}^{\infty} \log f_0(1 - \lambda e^{-i\alpha(\lambda_0-\varepsilon)}) \leq c \sum_{j=I(\varepsilon)}^{\infty} q_j g_j(\lambda),$$

where $g_j(\lambda) = \sum_{i=I(\varepsilon)}^{\infty} \left[1 - (1 - \lambda e^{-i\alpha(\lambda_0-\varepsilon)})^j\right]$. For each j, it is easy to show that $g_j(\lambda) \downarrow 0$ as $\lambda \to 0$. Also, we have already shown that if $\sum_{j=2}^{\infty} q_j \log j < \infty$, then $\sum_{j=2}^{\infty} q_j g_j(\lambda) < \infty$ for any $\lambda > 0$. Hence by the Monotone Convergence Theorem, $\lim_{\lambda \to 0} \sum_{j=1}^{\infty} q_j g_j(\lambda) = 0$, which proves that $\varphi(\lambda) \to 1, \lambda \to 0$.

When $\sum_{j=2}^{\infty} p_j j \log j < \infty$ it is not difficult to show that $\limsup_{|\lambda| \to \infty} |\psi(i\lambda)| = f(q) < 1$ and hence $\limsup_{|\lambda| \to \infty} |\varphi(i\lambda)| \leq [f_0(f(q))]^N$. Since N is arbitrary then $\limsup_{|\lambda| \to \infty} |\varphi(i\lambda)| = 0$, which completes the proof of Theorem 5.4 (Ib).

Assume now that $m_0 < \infty$. Slavtchova and Yanev (1990) have shown that $\sum_{j=2}^{\infty} j p_j \log j = \infty \Longleftrightarrow \psi(\lambda) \equiv 1$. Theorem 5.4 (IIc) follows immediately.

Let now $\sum_{j=2}^{\infty} j p_j \log j < \infty$. It follows from the proof of Theorem 5.4 (Ib) that $\varphi(\lambda)$ is the L.T. of a proper r.v. and from (5.1) – (5.3) it follows that $\mathbb{P}[W = 0] = \lim_{\lambda \to 0} \varphi(\lambda) = 0$.

To complete the proof we have to show the absolute continuity of the distribution of the r.v. W. Denote $\bar{\varphi}(t) = \mathbb{E}e^{itW}$ and $\bar{\psi}(t) = \mathbb{E}e^{it \lim[Z(t)e^{-\alpha t}]}$. Equation (5.1) yields $\bar{\varphi}'(t) = \bar{a}\mathbb{E}[\bar{\varphi}'(t\xi)f_0(\bar{\psi}(t\xi))\xi] + R(t)$, where $\bar{a} = \int_0^{\infty} e^{-\alpha t} dG_0(t)$, $\xi = e^{-\alpha \hat{U}}$, $G_0^{(\alpha)}(x) = \mathbb{P}(\hat{U} \leq x) = (1/\bar{a}) \int_0^x e^{-\alpha t} dG_0(t)$ and

$$R(t) = \bar{a} \int_0^{\infty} \bar{\varphi}(te^{-\alpha u})\bar{\psi}'(te^{-\alpha u})f_0(\bar{\psi}(te^{-\alpha u}))e^{-\alpha u} dG_0(u).$$

Let $J(T) = \int_{-T}^{T} |\bar{\varphi}'(t)| dt$. Using Theorem 2 (Athreya and Ney (1972), Chapter IV) and Corollary 4.4 (Slavtchova and Yanev (1991)), it is not difficult to show that $\int_{-T}^{T} |R(t)| dt \leq c$, where c is a positive constant. Then

$$J(T) \leq \bar{a} \int_{-T}^{T} \mathbb{E}[|\bar{\varphi}'(t\xi)\xi|] dt + \int_{-T}^{T} |R(t)| dt \leq \bar{a}\mathbb{E}[J(T\xi)] + c.$$

An iteration yields

$$J(T) \leq \bar{a}^n \mathbb{E}\left[J(T\prod_{i=0}^{n} \xi_i)\right] + c\{\bar{a}^{n-1} + \ldots + \bar{a} + 1\}.$$

Since $\prod_{i=1}^{n} \xi_i = exp\{-\alpha \sum_{i=1}^{n} \hat{U}_i\}$, then by the LLN we have $\prod_{i=1}^{n} \xi_i \to 0$ a.s. Hence $J(T) \leq c/(1-\bar{a})$. As $T \to \infty$ we obtain $\int_{-\infty}^{\infty} |\bar{\varphi}'(t)| dt < \infty$ and by Lemma 3 (see Athreya (1969)) it follows that W has an absolutely continuous distribution on $[0, \infty)$.

8   Proof of Theorem 5.5.

In the case $p_0 = 0$ it follows that $X(t)$ is a Bellman - Harris branching process with renewal type of immigration and it is known (see Kaplan and Pakes (1974)) that the assumptions of the theorem are sufficient for a.s. convergence.

Let $p_0 > 0$. By Slavtchova and Yanev (1991) and Jagers (1968) it follows that if $\sum_{i=1}^{\infty} j^2 p_j < \infty$, then $\lim_{t \to \infty} Z_{ij}(t)/\bar{A}e^{\alpha t} = \tilde{W}_{ij}$ a.s., whose L.T. satisfies (5.2). This implies in particular that for each i, $\lim_{t \to \infty} \sum_{j=1}^{\nu_i} Z_{ij}(t)/\bar{A}e^{\alpha t} = W_i$ a.s. The r.v. W_i has L.T. $f_0(\psi(u))$. It follows from the assumptions of Section 1 that $\{W_i\}$ are i.i.d. and are also independent of $\{\tau_i\}$. Define $W = \sum_{i=1}^{\infty} e^{-\alpha\tau_i} W_i$. To complete the theorem one has to show that $\int_0^{\infty} \mathbb{E}[W(t) - W]^2 dt < \infty$. Observe that

$$[W(t) - W]^2 \leq 2\left[(W(t) - \sum_{i=1}^{n(t)} e^{-\alpha\tau_i} W_i)^2 + (\sum_{i=n(t)+1}^{\infty} e^{-\alpha\tau_i} W_i)^2\right] = 2[R_1(t) + R_2(t)].$$

From (5.5) by Schwarz's inequality,

$$R_1(t) \leq \left[\sum_{i=1}^{n(t)} e^{-\alpha\tau_i}\right]\left[\sum_{i=1}^{n(t)} e^{-\alpha\tau_i} \left\{\sum_{j=1}^{\nu_i} Z_{ij}(t - \tau_i)/\bar{A}e^{\alpha(t-\tau_i)} - -W_i\right\}^2\right].$$

It is not difficult to show that

$$\int_0^\infty \left[\sum_{i=1}^{n(t)} e^{-\alpha \tau_i} \left\{ \sum_{j=1}^{\nu_i} Z_{ij}(t - \tau_i)/\bar{A}e^{\alpha(t-\tau_i)} - -W_i \right\} \right]^2 dt$$

$$= \sum_{i=1}^\infty e^{-\alpha \tau_i} \int_{\tau_i}^\infty \left\{ \sum_{j=1}^{\nu_i} Z_{ij}(t - \tau_i)/\bar{A}e^{\alpha(t-\tau_i)} - W_i \right\}^2 dt$$

$$= \sum_{i=1}^\infty e^{-\alpha \tau_i} \int_0^\infty \left\{ \sum_{j=1}^{\nu_i} Z_{ij}(t)/\bar{A}e^{\alpha t} - W_i \right\}^2 dt.$$

By independence we conclude that

$$\mathbb{E}\left[\int_0^\infty R_1(t)dt \right] \le \mathbb{E}\left\{ \left(\sum_{i=1}^\infty e^{-\alpha \tau_i} \right)^2 \right\} \mathbb{E}\left\{ \int_0^\infty \left[\sum_{j=1}^{\nu_1} (Z_{1j}(t)/\bar{A}e^{\alpha t} - \tilde{W}_{1j}) \right]^2 dt \right\}.$$

Note that $\mathbb{E}\left[\sum_{i=1}^\infty e^{-\alpha \tau_i} \right]^2 < \infty$. Also by Schwarz's inequality,

$$\mathbb{E}\left\{ \int_0^\infty \left[\sum_{j=1}^{\nu_1} (Z_{1j}(t)/\bar{A}e^{\alpha t} - \tilde{W}_{1j}) \right]^2 dt \right\} \le \mathbb{E}\{\nu_1^2\} \int_0^\infty \mathbb{E}[Z_{11}(t)/\bar{A}e^{\alpha t} - W_1]^2 dt.$$

However, Slavtchova and Yanev (1991) have shown that the last integral is finite if $\sum_{j=1}^\infty j^2 p_j < \infty$. Similarly

$$\mathbb{E}\left[\int_0^\infty R_2(t)dt \right] \le \mathbb{E}\left\{ \left(\sum_{i=0}^\infty e^{-\alpha \tau_i} \right) \left(\sum_{i=1}^\infty \tau_i e^{-\alpha \tau_i} W_i^2 \right) \right\} \mathbb{E}W_1^2$$

$$\le \delta \mathbb{E}\left(\sum_{i=1}^\infty e^{(-\alpha/2)\tau_i} \right)^2 \mathbb{E}W_1^2,$$

for $\delta > 0$ such that $xe^{-\alpha x} \le \delta e^{-\alpha x/2}$, $x > 0$. On the other hand,
$\mathbb{E}W_1^2 = \mathbb{E}\left(\sum_{j=1}^{\nu_1} \tilde{W}_{1j} \right)^2 \le \mathbb{E}\{\nu_1^2\}\mathbb{E}\{\tilde{W}_{11}^2\}$. The assumption $\sum_{j=1}^\infty p_j j^2 < \infty$ implies
$\mathbb{E}\{\tilde{W}_{11}^2\} < \infty$. Therefore $\mathbb{E}\int_0^\infty R_2(t)dt < \infty$, which completes the proof.

Acknowledgements. The authors are very grateful to the referee for useful comments and suggestions.

References

Abramowitz, M. and Stegun, I. A.(1970) *Handbook of Math. Functions with Formulas, Graphs and Math. Tables.* Dover, New York.

Athreya, K. (1969) On the supercritical one dimentional age- dependent branching processes.*Ann. Math.Statist.* **40,** 743 - 763.

Athreya, K. and Ney, P.(1972) *Branching Processes.* Springer Verlag, Berlin.

Foster, J.H. (1971) A limit theorems for a branching process with state-dependent immigration. *Ann. Math. Statist.* **42,** 1773-1776.

Harris, T. (1963) *The Theory of Branching Processes.* Springer Verlag, Berlin.

Jagers, P. (1968) Renewal theory and almost sure convergence of branching processes. *Ark. Mat.* **7,** 495-504.

Jagers, P. (1968) Age-dependent branching processes allowing immigration. *Theory Probab. Appl.* **13,** 225-236.

Kaplan, N. and Pakes A. (1974) Supercritical age-dependent branching processes with immigration. *Stoch. Proc. Appl.* **2,** 371-389.

Mitov, K. V. and Yanev N. M. (1985) Bellman-Harris Branching processes with state-dependent immigration. *J. Appl. Prob.* **22,** 757-765.

Mitov, K. V. and Yanev N. M. (1989) Bellman-Harris Branching processes with a special type of state-dependent immigration. *Adv. Appl. Prob.* **21,** 270 - 283.

Pakes, A. G. (1971) A branching processes with a state-dependent immigration component.*Adv. Appl. Prob.* **3,** 301-314.

Pakes, A. G. (1975) Some results for non-supercritical Galton-Watson processes with immigration. *Math. Biosci.* **24,** 71 - 92.

Sevastyanov, B. A. (1957) Limit theorems for branching processes of a special type.*Theory Prob. Appl.* **2,** 339 - 348.

Slavtchova, M. and Yanev N. (1990) Convergence in distribution of supercritical Bellman-Harris branching processes with state-dependent immigration. *Mathematica Balkanica, New series* **4,** 1, 35-42.

Slavtchova, M. and Yanev, N. (1991) Non - Critical Bellman - Harris branching processes with state-dependent immigration. *Serdica* **17,** 67-79.

Slavtchova-Bojkova, M. and Yanev, N. (1994) Limit theorems for age-dependent branching processes with state-dependent immigration. *Preprint, Institute of Mathematics, Bulgarian Academy of Sciences.*

Weiner, H. (1991) Age-dependent branching processes with two types of immigration. *Journal of Information and Optimization Sciences* **2,** 207-218.

Yamazato, M. (1975) Some results on continuous time branching processes with state-dependent immigration.*J. Math. Soc. Japan* **27,** 479 - 496.

Yanev, N. M. (1972) On a class of decomposed age-dependent branching processes.*Mathematica Balkanica* **2,** 58 - 75.

A NEW CLASS OF BRANCHING PROCESSES

S.R. ADKE * AND V. G. GADAG †

Abstract

A unified formulation to generate branching processes with continuous or discrete state space is provided. It includes processes with immigration and in varying environments. It also expands the known class of non-Gaussian Markov time series for non-negative variates. The finite time and asymptotic properties of the processes introduced in the paper are investigated.

Key words: Branching process in varying environments, Immigration, Non-gaussian time series, Limit distributions, Stationary gamma and Negative bionmial branching processes.
AMS 1991 Subject Classification: Primary 60J80, Secondary 60J05.

1   Introduction

Let $\{W_{in}, i \geq 1, n \geq 1\}$ be a double array of independent and identically distributed (i.i.d.) non-negative integer-valued random variables (r.v.'s) called the offspring r.v.'s. Let $\{U_n, n \geq 1\}$ be a sequence of i.i.d. non-negative integer-valued r.v.'s called the immigration r.v.'s, which are assumed to be independent of W_{in}'s. A simple model to describe the fluctuations in the sizes X_n, $n = 0, 1, \ldots$ of the successive generations of a population of individuals due to births, deaths and immigration was introduced by Heathcote (1965, 66). He defined X_n's recursively by the relation

$$X_{n+1} = \sum_{i}^{X_n} W_{in+1} + U_{n+1}, \quad n = 0, 1, 2, \ldots \qquad (1.1)$$

where the initial population size X_0 is an arbitrary non-negative integer-valued r.v., independent of W_{in}'s and U_n's.

Later, Kallenberg (1979) introduced a branching process (BP) with continuous state-space to model situations when it is difficult to count the number of individuals in the population but a related non-negative variable like volume, weight or toxin produced by the "individuals" of the population can be measured. In this paper we combine these two features of the existing models and introduce a generalized BP with immigration and in

---

*Postal address: Department of Statistics, University of Poona, Pune - 411 007, India.
†Postal address: Division of Community Medicine and Department of Mathematics and Statistics, Memorial University of Newfoundland, St. John's, Newfoundland, A1B 3V6.

varying environments, which can accommodate both discrete and continuous state-spaces. We make the following assumptions.

(A.1) Let $\mathbf{W}_n = \{W_{in}, \ i \geq 1\}$, be a sequence of i.i.d. non-negative r.v.'s, each W_{in} begin distributed like W_n, say. The double array $\mathbf{W} = \{\mathbf{W}_n, \ n \geq 1\}$ consists of independent sequences $\mathbf{W}_n, \ n = 1, 2, \dots$.

(A.2) Each of the stochastic processes $\mathbf{N}_n = \{N_n(t), \ t \epsilon T\}, \ n = 1, 2, \dots$ has state-space Z^+, the set of non-negative integers. They are independent processes and each $\mathbf{N}_n$ has stationary and independent increment (s.i.i.) with $N_n(0) = 0$ almost surely (a.s.). Here T is either the non-negative half $[0, \infty)$ of the real line $\mathbf{R}$ or Z^+.

(A.3) The sequence $\mathbf{U} = \{U_n, \ n \geq 1\}$ consists of independent non-negative r.v.'s U_n.

(A.4) The $\mathbf{N} = \{\mathbf{N}_n, \ n \geq 1\}$, $\mathbf{W}$ and $\mathbf{U}$ processes are independent.

(A.5) The initial r.v. X_0 is an arbitrary, non-negative r.v., independent of all other processes introduced in A.1 - A.4.

The main object of study in this paper is the process $\mathcal{X} = \{X_n, \ n \geq 0\}$ defined by the recursive relation

$$X_{n+1} = \sum_{i=1}^{N_{n+1}(X_n)} W_{in+1} + U_{n+1}, \quad n = 0, 1, 2, \dots \tag{1.2}$$

In what follows we refer to W_{in}'s and U_n's as offspring and immigration r.v.'s respectively of the n-th generation, and the $\mathbf{N}$ as the process which governs or derives the $\mathcal{X}$ - process.

The assumption A.1 retains the basic feature of a BP that the offspring distribution is the same for all 'individuals' of a generation. However, by allowing it to vary over generations, we ensure that the $\mathcal{X}$ - process is a BP with immigration and in varying environments.

In A.2, it is necessary to assume that each $\mathbf{N}_n$ has state-space Z^+ to ensure the validity of the sum on the right of (1.2). Thus one may replace the s.i.i. process $\mathbf{N}_n$ by any other process with state-space Z^+. However, the s.i.i. property of $\mathbf{N}_n$ implies that the $\mathcal{X}$ - process has properties similar to those of the Heathcote model (1.1). The case $T = \mathbf{R}^+$ suffices for most of the situations. We allow $T = Z^+$ to accommodate the Heathcote model as a particular case of (1.2). The assumptions A.3 - A.5 are of a technical nature.

It is worth noting that the Kallenberg (1979) approach can accommodate neither immigration nor varying environments, which is possible with our formulation of the model. We also do not require the infinite divisibility of the offspring distribution as is required by Kallenberg (1979) for the existence of his process.

The $\mathcal{X}$-process defined by (1.2), induces the $\mathbf{N}(\mathcal{X}) = \{N_{n+1}(X_n); \ n \geq 0\}$ - process. The $\mathcal{X}$ - and $\mathbf{N}(\mathcal{X})$ - processes provide a generalization of the approaches of Mckenzie

(1986), Al-Osh and Alzaid (1987), Sim (1990), Adke and Balakrishna (1992) and Al-Osh and Aly (1992), for generating non-Gaussian stationary Markov sequences with specified one-dimensional marginal distributions.

The processes obtained from (1.2) when $U_n \equiv 0$ a.s., i.e. when immigration is absent are also of interest. We denote by $\mathcal{X}^*$ the process $\{X_n^*, n \geq 1\}$ where

$$X_{n+1}^* = \sum_{i}^{N_{n+1}(X_n^*)} W_{in+1}, \; n = 0, 1, \dots \tag{1.3}$$

and by $\mathbf{N}(\mathcal{X}^*)$ the process $\{N_{n+1}(X_n^*), n \geq 0\}$. These processes have interesting interpretations in relation to a single sever queue in which the arrival and service times are independent and the queue discipline is the usual first come, first served. Interpret X_0^* as the service time of the initial customer and $N_1(X_0^*)$ as the number of customers joining the queue during $(0, X_1^*]$. Then X_1^* is the total service time of these $N_1(X_0^*)$ customers. More generally, X_{n+1}^* is the total service time of the $N_{n+1}(X_n^*)$ customers joining the queue during $(X_0^* + \cdots + X_{n-1}^*, X_0^* + \cdots + X_n^*]$. The server becomes free after serving n batches, if $N_{n+1}(X_n^*)$ is zero. Thus one may interpret $\sum_{j=0}^{\infty} X_j^*$ as the busy period of the server initiated by the initial customer at a queue in which the arrivals are governed by the $\mathbf{N}_n$ - processes and W_{in}'s are the service times.

The organization of the paper is as follows. The distributional properties of the $\mathcal{X}, \mathcal{X}^*, \mathbf{N}(\mathcal{X})$ and $\mathbf{N}(\mathcal{X}^*)$ processes are described in section 2. A method for obtaining the probabilities of extinction of the $\mathcal{X}^*$ and $\mathbf{N}(\mathcal{X}^*)$ processes is provided in section 3. The asymptotic properties like existence of the stationary distribution for the $\mathcal{X}$-process are obtained in section 4. The results of sections 2, 3, and 4 are illustrated with examples in section 5. These examples lead to a problem of identification which is also discussed in section 5. The versatality of our model and directions of future work are briefly discussed in the last section, section 6.

2 Distributional Properties

The sequence $\{X_n, n \geq 0\}$ is by definition a non-homogeneous Markov sequence with state-space S, say. Its distributional properties can be studied in terms of the Laplace transforms (L.T.'s) introduced below. Let $s \epsilon \mathbf{R}^+$ and define for $n \geq 1$

$$F_n(s) = E\{\exp(-sN_n(1))\}, \quad f_n(s) = -\log F_n(s)$$

$$G_n(s) = E\{\exp(-sW_n)\}, \quad g_n(s) = -\log G_n(s) \tag{2.1}$$

and

$$H_n(s) = E\{\exp(-sU_n)\}, \quad h_n(s) = -\log H_n(s).$$

In view of the s.i.i. property of the $\mathbf{N}_n$ - process we have

$$F_n(s,t) = E\{\exp(-sN_n(t))\} = \{F_n(s)\}^t \tag{2.2},$$

and therefore,

$$f_n(s,t) = -\log F_n(s,t) = t f_n(s), \quad t \epsilon T. \tag{2.3}$$

Lemma 2.1. The L.T. $L_n(s) = E\{\exp(-sX_n)\}$ satisfies the recurrence relation

$$L_{n+1}(s) = H_{n+1}(s)L_n(a_{n+1}(s)), \tag{2.4}$$

where $a_{n+1}(s) = f_{n+1}(g_{n+1}(s))$.

Proof: Observe that by virtue of (1.2), (2.2) and Assumptions A.1 - A.4,

$$K_{n+1}(x,s) = E\{\exp(-sX_{n+1})|X_n = x\}$$

$$= H_{n+1}(s)\{F_{n+1}(g_{n+1}(s))\}^x, \quad n \geq 0, \; x\epsilon S. \tag{2.5}$$

Hence, by assumption A.4

$$L_{n+1}(s) = EE\{\exp(-sX_{n+1})|X_n\}$$

$$= H_{n+1}(s)E\{(F_{n+1}(g_{n+1}(s)))^{X_n}\}$$

$$= H_{n+1}(s)L_n(a_{n+1}(s))$$

which is (2.4).

Remark 2.1. The L.T. $K_{n+1}^*(x,s)$ of the transition kernel $P[X_{n+1}^* \leq y|X_n^* = x]$ and the L.T. $L_{n+1}^*(x)$ of X_{n+1}^* are obtainable from (2.5) and Lemma 2.1 by putting $H_{n+1}(s) \equiv 1$.

Remark 2.2. The relation (2.5) gives the recursive relation satisfied by the L.T. of the transition kernels of the non-homogeneous Markov sequence (1.2).

In what follows, the symbols for the descriptors for the $\mathcal{X}^*$-process are same as those of the $\mathcal{X}$-process with a superscript (*) imposed on them.

The following theorem describes the structure of the sequence $\{N_{n+1}(X_n), n \geq 0\}$ which governs the evolution of the $\mathcal{X}$-process.

Theorem 2.1. The sequence $\{N_{n+1}(X_n), n \geq 0\}$ is a Galton - Watson process with immigration (GWI process) in varying environments.

Proof: Let $n \geq 1$. Observe that if $N_n(X_{n-1}) = 0$,

$$N_{n+1}(X_n) = N_{n+1}(U_n).$$

When $N_n(X_{n-1}) > 0$,

$$N_{n+1}(X_n) = \sum_{i=1}^{N_n(X_{n-1})} \{N_{n+1}(\sum_{j=1}^{i} W_{jn}) - N_{n+1}(\sum_{j=1}^{i-1} W_{jn})\}$$

$$+ \{N_{n+1}(\sum_{i=1}^{N_n(X_{n-1})} W_{in} + U_n) - N_{n+1}(\sum_{i=1}^{N_n(X_{n-1})} W_{in})\}$$

$$= \sum_{i=1}^{N_n(X_{n-1})} \xi_{in+1} + \eta_{n+1} \text{ , say.} \tag{2.6}$$

In view of the s.i.i. property of the $\mathbf{N}_{n+1}$ process, it is easy to recognize that the r.v.s. ξ_{in+1}, $i \geq 1$ and η_{n+1} are independently distributed. Further, each ξ_{in+1}, $i \geq 1$, has the same distribution as that of $N_{n+1}(W_n)$. The distribution of η_{n+1} is same as that of $N_{n+1}(U_n)$. Moreover, the r.v.s. ξ_{in+1}, $i \geq 1$ and η_{n+1} are independent of $N_n(X_{n-1})$ by construction.

Thus it follows from (2.6) that $\{N_{n+1}(X_n),\ n \geq 0\}$ is a GWI process in varying environments. Its initial distribution has the L.T.

$$E\{\exp(-sN_1(X_0))\} = L_0(f_1(s)). \tag{2.7}$$

The L.T.'s of the of the offspring and the immigration distributions for the $(n+1)$-th generation are

$$E\{\exp(-sN_{n+1}(W_n))\} = G_n(f_{n+1}(s));$$
$$E\{\exp(-sN_{n+1}(U_n)\} = H_n(f_{n+1}(s)); \tag{2.8}$$

respectively. The proof is complete.

Corollary 2.1. The L.T. $R_n(s) = E\{\exp(-sN_n(X_{n-1}))\}$ satisfies the recurrence relation

$$R_{n+1}(s) = H_n(f_{n+1}(s))\, R_n(b_n(s)),\ n \geq 1 \tag{2.9}$$

where $b_n(s) = g_n(f_{n+1}(s))$ and $R_1(s) = L_0(f_1(s))$.

Proof: This is a consequence of (2.6), (2.7) and (2.8).

Corollary 2.2. The sequence $\{N_{n+1}(X_n^*),\ n \geq 0\}$ is a Bienayme - Galton - Watson (BGW) process in varying environments, cf. Jagers (1975).

The first and second order moments $m_n = E(X_n))$, $\sigma_n^2 = Var(X_n)$ and $\sigma_{mn} = Cov(X_m, X_n)$, $m < n$, $n \geq 0$, associated with the $\mathcal{X}$ process are specified in terms of

$$E(W_n) = \mu_n \text{ , } Var(W_n) = \gamma_n^2,$$
$$E(N_n(1)) = \lambda_n \text{ , } Var(N_n(1)) = \delta_n^2, \tag{2.10}$$
$$\text{and} E(U_n) = \nu_n,\ Var(U_n) = \tau_n^2,\ n \geq 1,$$

which are all assumed to be finite.

Lemma 2.2. We have for $n \geq 1$ and $n > m$

$$m_n = \nu_n + \theta_n \, m_{n-1} = \Theta_n \{ m_0 + (\sum_{j=1}^{n} \Theta_j^{-1} \nu_j) \}, \tag{2.11}$$

$$\sigma_n^2 = \tau_n^2 + (\lambda_n \gamma_n^2 + \delta_n^2 \mu_n^2) \, m_{n-1} + \theta_n^2 \sigma_{n-1}^2 \tag{2.12}$$

$$= \Theta_n^2 \{ \sigma_0^2 + \sum_{j=1}^{n} \Theta_j^{-2} (\tau_j^2 + (\lambda_j \gamma_j^2 + \delta_j^2 \mu_j^2) \, m_{j-1}) \}$$

$$\sigma_{mn} = \Theta_n \Theta_m^{-1} \sigma_m^2 \tag{2.13}$$

where

$$\theta_n = \lambda_n \mu_n \tag{2.14}$$

and $\Theta_n = \Pi_{i=1}^{n} \theta_j$.

Proof: The above results are obtainable from (1.2).

The corresponding results for the $\mathcal{X}^*$ - process are obtained by putting $\mu = \tau^2 = 0$ in the above relations. The second order moments of the GWI $\{N_n(X_{n-1}), \, n \geq 1\}$ as well as of the BP $\{N_n(X_{n-1}^*), \, n \geq 1\}$ can be obtained using (2.2) and the usual conditioning techniques.

The varying nature of the environments of the $\mathcal{X}$ and the $\mathbf{N}(\mathcal{X})$ - processes is a consequence of the non-identical distributions of the components of the $\mathbf{N}, U$ and $\mathbf{W}$ processes. Thus, in order to obtain the generalization of the classical GWI process (1.1) in fixed enviroments, we introduce the following additional assumption.

A.6. The components $\mathbf{N}_n, \, n \geq 1$ of the $\mathbf{N}$ process are identically distributed and so are the components of $\mathbf{W}$ as well as of $\{U_n, \, n \geq 1\}$.

The above assumption A.6. allows us to drop the suffix n of $\mu_n, \gamma_n^2, \lambda_n, \delta_n^2, \nu_n, \tau_n^2$, the Laplace transforms introduced in (2.1), and from all other appropriate functions.

Corollary 2.3. Let $n \geq 1$ and let A.6 hold. When $\theta = E\{N(1)\} \, E(W) = \lambda \mu \neq 1$,

$$m_n = \nu \{ (1 - \theta^n)/(1 - \theta) \} + \theta^n m_0 \tag{2.15}$$

and

$$\sigma_n^2 = \tau^2 (1 - \theta^{2n})/(1 - \theta^2) \tag{2.16}$$

$$+ m_0 \{ \lambda \gamma^2 + \delta^2 \mu^2 \} \theta^{n-1} (1 - \theta^n)/(1 - \theta)$$

$$+ \nu \{ \lambda \gamma^2 + \delta^2 \mu^2 \} (1 - \theta^n)(1 - \theta^{n-1})/\{ (1 - \theta)(1 - \theta^2) \}$$

$$+ \theta^{2n} \sigma_0^2.$$

The appropriate modifications can be made when $\theta = 1$.

3   Extinction Probability

In this section we provide a method for obtaining the probabilities $q_n = P[X_n^* = 0]$ and $\rho_n = P[N_n(X_{n-1}^*) = 0]$ of extinction at or before the n-th generation of the $\mathcal{X}^*$ and $\mathbf{N}(\mathcal{X}^*)$ processes respectively. If $P[W_n = 0] = 0$, it follows by the definition of the $\mathcal{X}^*$-process that $q_n = \rho_n$, $n \geq 1$. However, if $P[W_n = 0] > 0$.

$$q_n = \rho_n + P[N_n(X_{n-1}^*) > 0, \ W_{1n} = \cdots = W_{N_n(X_{n-1}^*)n} = 0]. \tag{3.1}$$

Lemma 3.1. The extinction probability q_n satisfies the relation

$$q_n = R_n^*(g_n(\infty)) \tag{3.2}$$

where $g_n(\infty) = \lim_{s \to \infty} g_n(s) = -\log P[W_n = 0]$.

Proof: Observe that from (3.1), we get

$$q_n = \sum_{j=0}^{\infty} P[N_n(X_{n-1}^*) = j]\{P[W_n = 0]\}^j \tag{3.3}$$

$$= E[\{P[W_n = 0]\}^{N_n(X_{n-1}^*)}]$$

which yields (3.2).

Since $[X_n^* = 0] \subset [X_{n+1}^* = 0]$, $q_n \leq q_{n+1} \leq 1$ and therefore, $q = \lim q_n$ exists, $0 < q \leq 1$, without any additional conditions. A similar statement may be made about the sequence $\{\rho_n : n \geq 1\}$. The relation (3.1) implies that $q \geq \rho = \lim \rho_n$. If $\rho = 1$, cf. Corollary 3.5.4 of Jagers (1975), p. 71, then $q = 1$. If $\rho < 1$, $q\epsilon[\rho, 1]$. The problem of locating $q\epsilon[\rho, 1]$ needs further study. A partial answer to this problem is provided in the following lemma.

Let N_∞^*, with values in $Z^+ \cup \{\infty\}$, be the almost sure limit of $N_n(X_{n-1}^*)$ as $n \to \infty$. The existence of this limit is guaranteed by theorem 3.5.1 of Jagers (1975), p. 70 and our corollary 2.2. Obviously $\rho = P[N_\infty^* = 0]$.

Lemma 3.2. Suppose $\alpha_n = P[W_n = 0] \to \alpha$ as $n \to \infty$. Then

$$\rho \leq q = \sum_{j=0}^{\infty} \alpha^j P[N_\infty^* = j]. \tag{3.4}$$

Proof: This is a consequence of (3.3) and the corollary 1 or Loève (1977), p. 205.

Remark 3.1. It is clear that $q = \rho$ if $\alpha = 0$ and $q = P[N_\infty^* < \infty] \leq 1$, if $\alpha = 1$. Further, $q\epsilon(0,1)$ if $\alpha\epsilon(0,1)$.

We discuss below the simplification obtained under the homogeneity assumption A.6, valid for the rest of this section. In order to highlight the similarity with Galton-Watson processes, we discuss the properties of the probability $q(1)$ of extinction of the $\mathcal{X}^*$-process initiated by the r.v. X_0^* degenerate at 1.

Lemma 3.3. If $X_0^* = 1$ a.s. then $k = -\log q(1)$ is the maximal solution in $[0, \infty)$ of the equation

$$s = a(s) , \tag{3.5}$$

where $a(s) = f(g(s))$.

Proof: By remark 2.1,

$$-\log E(e^{-sX_{n+1}^*}) = \ell_{n+1}^*(s) = \ell_n^*(a(s))$$

which on iteration yields

$$\ell_{n+1}^*(s) = a_{(n+1)}(s) = a(a_{(n)}(s)), \tag{3.6}$$

using the fact that $\ell_0^*(s) = s$. In the above $a_{(n)}(.)$ is the n-th functional iterate of $a(s)$. Allow $s \to \infty$ in (3.6) to verify that

$$k_{n+1} = a(k_n) \tag{3.7}$$

where $k_n = -\log q_n(1)$. If in the equation (3.7) we allow $n \to \infty$, we find that k is a solution of (3.5). The fact that k is the maximal solution of (3.5) follows as in theorem 3.3.2 of Kallenberg (1979), p. 18.

Corollary 3.1. If X_0^* is an arbitrary non-negative r.v., the extinction probability

$$q = E[\{q(1)\}^{X_0^*}].$$

Similar results can be stated for $\mathbf{N}(\mathcal{X}^*)$ process. It is interesting to note that

$$E[X_n^* | X_{n-1}^* = 1] = \theta = \lambda\mu = E[N_{n+1}(X_n^*) | N_n(X_{n-1}^*) = 1].$$

Thus, the criticality parameter θ is the same for the $\mathcal{X}^*$ - and $\mathbf{N}(\mathcal{X}^*)$-processes.

4   Limit Distributions

The following results give the analogues of the classical limit theorems guarenteeing existence of the a.s. limits in Galton-Watson processes under appropriate normings. By virtue of Lemma 2.2,

$$E(N_{n+1}(X_n)) = \lambda_{n+1}E(X_n)$$

$$= \lambda_{n+1}m_n = \lambda_{n+1}\Theta_n\{m_0 + \sum_{j=1}^{n}\Theta_j^{-1}\nu_j\}$$

Thus, it is reasonable to consider the limit behaviour of $V_n = \Theta_n^{-1}X_n$ and of $Y_n = (\lambda_{n+1}\Theta_n)^{-1}N_{n+1}(X_n)$, as $n \to \infty$.

Theorem 4.1. If $\sum\limits_{n=1}^{\infty} \Theta_n^{-1}\nu_n < \infty$, the sequence $\{V_n; n \geq 1\}$ converges a.s. to a non-negative r.v. V.

Proof: Using (2.11), it is easy to verify that $\{V_n; n \geq 1\}$ is a non-negative sub-martingale and that

$$E(V_n) = m_0 + \sum_{j=1}^{n}\Theta_j^{-1}\nu_j. \tag{4.1}$$

Thus, under the hypothesis of the theorem $\sup\limits_{n} E(V_n) < \infty$. Hence, the theorem follows by the sub-martingale convergence theorem.

Corollary 4.1. Let $\sum\limits_{n=1}^{\infty} \Theta_n^{-1}\nu_n < \infty$. Then there exists a non-negative r.v. Y such that $Y_n \to Y$ almost surely.

Corollary 4.2. Let $V_n^* = \Theta_n^{-1}X_n^*$ and $Y_n^* = (\lambda_{n+1}\Theta_n)^{-1}N_{n+1}(X_n^*)$. There exists non-negative r.v.s. V^* and Y^* such that $V_n^* \to V^*$ and $Y_n^* \to Y^*$ both with probability one.

In view of the possible applications of our model to the generation of stationary non-Gaussian Markov time series of non-negative r.v.s, it is of interest to study the existence of limit distributions for the $\mathcal{X}$ and $N(\mathcal{X})$-processes under the homogeneity assumption A.6.

Remark 4.1. When A.6 holds, the $N(\mathcal{X})$-process is a GWI process by virtue of theorem 2.1. The cumulant generating function (c.g.f.) of the corresponding offspring distribution is

$$b(s) = g(f(s)). \tag{4.2}$$

The c.g.f. of the immigration r.v. $N(U)$ is

$$\psi(s) = h(f(s)). \tag{4.3}$$

The following well-known result (cf. Jagers (1975), p. 55, 56) is stated here for ready reference.

Theorem 4.2. Let A.6 hold.

(i) If $\theta = \lambda\mu < 1$, then as $n \to \infty$, $N_n(X_{n-1})$ converges in distribution to a limit r.v. N_∞, say. The c.g.f. $r_\infty(s)$ of N_∞ satisfies the equation

$$r_\infty(s) = \psi(r_\infty(b(s))). \tag{4.4}$$

The limit r.v. N_∞ is honest if and only if

$$E[\log N(U)]^+ < \infty \tag{4.5}$$

(ii) If $\theta = 1$, ν, δ^2, and γ^2 are positive and fintie, then as $n \to \infty$, $[N_n(X_{n-1})/n]$ converges in distribution to a r.v. Z which has gamma $G(\alpha, \beta)$ distribution with density function.

$$g(z, \alpha, \beta) = [\beta^\alpha/\Gamma(\alpha)]z^{\alpha-1}\exp(-\beta z), \quad z\epsilon[0, \infty),$$

where $\alpha = 2\nu/(\delta^2\mu + \gamma^2\lambda^2)$ and $\beta = 2/(\delta^2\mu + \gamma^2\lambda^2)$.

The above theorem leads to the following corresponding results for the $\mathcal{X}$-process.

Theorem 4.3. Let A.6 hold and $\theta < 1$. Then as $n \to \infty$, X_n converges in distribution to X say, whose L.T. is

$$\Phi(s) = E[\exp(-sX)] = H(s)\ R_\infty(g(s)). \tag{4.6}$$

In the above $R_\infty(s) = \exp[-r_\infty(s)]$. The limit r.v. X is honest whenever (4.5) holds.

Proof: The result follows from theorem 4.2 (i) and the fact that

$$E[\exp(-sX_{n+1})] = H(s)\ E[\{G(s)\}^{N_{n+1}(X_n)}].$$

Theorem 4.4. Under the conditions of theorem 4.2 (ii), $n^{-1}X_n$ converges in distribution to a gamma $G(\alpha, \beta/\mu)$ r.v.

Proof: We have

$$E[\exp(-sX_n/n)] = H(s/n)\ E[\{G(s/n)\}^{N_n(X_{n-1})}]$$

$$= H(s/n)\ E[\{G^n(s/n)\}^{N_n(X_{n-1})/n}]$$

$$= \quad H(s/n) \, E[\{1 - s\mu/n + 0(1/n)\}^{nN_n(X_{n-1})/n}]$$

$$\rightarrow \quad E[\exp(-s\mu Z)] \quad \text{as} \quad n \rightarrow \infty.$$

by virtue of theorem 4.2 (ii). The proof is complete.

5 Examples

The results of the previous sections are illustrated with the help of the following examples

Example 5.1. Adke and Balakrishna (1992) have studied the properties of the $\mathcal{X}$-process, when $\mathbf{N}_n$ is a homogeneous Poisson process with rate λ_n, W_n has exponential distribution $\exp(\alpha_n)$ with mean α_n^{-1}, and U_n has $G(\nu_n, \alpha_n)$ distribution, $n \geq 1$. The initial r.v. X_0 has $G(\nu_0, \alpha_0)$ distribution. In this model, relation (2.5) simplifies to

$$(5.1) \qquad L_{n+1}(s) = \left[\frac{\alpha_{n+1}}{\alpha_{n+1} + s} \right]^{\nu_0} L_n \left(\frac{\lambda_{n+1} s}{\alpha_{n+1} + s} \right),$$

a consequence of which is that X_n has $G(\nu_0, M_n)$ distribution, where

$$(5.2) \qquad M_0 = \alpha_0, \quad M_n^{-1} = \alpha_n^{-1} + \lambda_n \alpha_n^{-1} M_{n-1}^{-1} \, , \, n \geq 1.$$

Thus, we may view the $\mathcal{X}$-process as a gamma BP with immigration in varying environments.

Adke and Balakrishna (1992) have not investigated the properties of the induced $\mathbf{N}(\mathcal{X})$-process.

Since X_{n-1} has $G(\nu_0, M_{n-1})$ distribution and $\mathbf{N}_n$ is a Poisson process, $N_n(X_{n-1})$ has negative binomial $NB(\nu_0, p_n)$ distribution with L.T.

$$R_n(s) = [\bar{p}_n/\{1 - p_n \exp(-s)\}]^{\nu_0};$$

where $\bar{p}_n = M_{n-1}/(M_{n-1} + \lambda_n) = 1 - p_n$. Consequently, the $\mathbf{N}(\mathcal{X})$ process is a negative binomial GWI process in varying environments. Its n-th generation offspring distribution is geometric, i.e. $NB(1, \lambda_{n+1}/(\alpha_n + \lambda_{n+1}))$. The corresponding immigration distribution is $NB(\nu_0, \lambda_{n+1}/(\alpha_n + \lambda_{n+1}))$. Recently, Al-Osh and Aly (1992) have given a method of constructing a stationary auto-regressive Markov chain with negative binomial and geometric marginals. These are the discrete analogues of the gamma and exponential processes of Sim (1990). The above discussion makes explicit the connection between the gamma auto-regressive processes of Sim and the negative binomial auto-regressive processes of Al-Osh and Aly. It also clarifies the nature of the sufficient conditions under which the $\mathcal{X}$- and $N(\mathcal{X})$-processes are stationary Markov processes.

In example 5.3, we give one more method of generating a Markov chain with negative binomial marginals.

In the absence of the immigration component, i.e., when $U_n \equiv 0$ a.s., we have from (5.1),

$$L_n^*(s) = L_{n-1}^*(s/(\theta_n^{-1} + \lambda_n^{-1}s)) \tag{5.3}$$

which, on iteration, gives

$$L_n^*(s) = L_0^*(s/(\Theta_n^{-1} + c_n^{-1}s)) $$

where $\Theta_0 = 1$ and $c_n^{-1} = \sum_{j=1}^{n} \lambda_j^{-1} \Theta_{j-1}^{-1}$. Thus, if $X_0^* = 1$ a.s.,

$$L_n^*(s) = \exp\{-c_n(1 - \Theta_n^{-1}c_n/(\Theta_n^{-1}c_n + s))\} \tag{5.4}$$

which implies that X_n^* is distributed like sum of i.i.d. $\exp(\Theta_n^{-1}c_n)$ r.v.s, the number of summands being an independent Poisson r.v. with mean c_n. An additional consequence of (5.4) is that

$$R_n^*(s) = \exp\{-c_{n-1}(1 - \bar{d}_n/(1 - d_n e^{-s}))\} \tag{5.5}$$

where

$$\bar{d}_n = \Theta_{n-1}^{-1}/(\Theta_{n-1}^{-1} + \lambda_n c_{n-1}^{-1}) = (1 - d_n). \tag{5.6}$$

Thus, $N_n(X_{n-1}^*)$ is distributed like the sum of i.i.d. geometric $NB(1, d_n)$ r.v.s, the number of summands being an independent Poisson r.v. with mean c_{n-1}.

Since W_n's do not have an atom at zero, the probabilities q_n and ρ_n of extinction by n-th generation of the $\mathcal{X}^*$ - and $\mathbf{N}(\mathcal{X}^*)$ - processes respectively, are the same. An explicit expression for q_n is obtained by allowing $s \to \infty$ in (5.4). In fact, when $X_0^* = 1$ a.s.

$$q_n = \lim_{s \to \infty} L_n^*(s) = \exp(-c_n),$$

and $q_n \to 1$ if and only if $c_n \to 0$, as $n \to \infty$. If $c_n \to c > 0$, $q_n = \exp(-c) < 1$. These results are easily extended to the case when X_0^* is an arbitrary positive r.v.

We now consider the homogeneous case obtaining when $\alpha_n \equiv \alpha$, $\lambda_n \equiv \lambda$, $n \geq 1$. The r.v. X_n has $G(\nu_0, M_n)$ distribution where

$$
M_n^{-1} = \begin{cases} \{\theta^n \alpha_0^{-1}\} + (1 - \theta^n)\{\alpha(1 - \theta)\}^{-1} & , \text{ when } \theta \neq 1 \\[2mm] \alpha_0^{-1} + n\alpha^{-1} & , \text{ when } \theta = 1. \end{cases}
$$

Obviously, the $\mathcal{X}$-process has a limit distribution which is $G(\nu_0, \alpha(1 - \theta))$ if and only if $\theta = \lambda/\alpha < 1$, a case discussed by Sim (1990). Thus, when $\theta < 1$ and $\alpha_0 = \alpha(1 - \theta)$, the $\mathcal{X}$-process is a stationary gamma BP with immigration. The induced $\mathbf{N}(\mathcal{X})$-process is a stationary negative binomial GWI process, with $N_n(X_{n-1})$ having $NB(\nu_0, (1 - \theta))$ distribution.

When A.6 holds and there is no immigration and $X_0^* = 1$, $c_n^{-1} = (\theta^n - 1)/\{\lambda\theta^{n-1}(\theta - 1)\}$ if $\theta \neq 1$ and $c_n^{-1} = n/\lambda$ if $\theta = 1$. The probability of ultimate extinction is 1 and if $\theta \leq 1$ and is $\exp(-(\lambda - \alpha))$ if $\theta > 1$ i.e. if $\lambda > \alpha$. These results are in agreement with theorem 3.5.2 and corollary 3.5.4 of Jagers (1975) p. 70-71.

Example 5.2. In this example we assume that $\mathbf{N}_n$-process is the compound Poisson process defined by

$$
N_n(t) = \sum_{\epsilon}^{A_n(t)} \psi_{in}
$$

where ψ_{in}'s are i.i.d. geometric $NB(1, \bar{p}_n)$ r.v.s, $0 < \bar{p}_n < 1$ and $\{A_n(t), t\epsilon \mathbf{R}^+\}$ is an independent homogeneous Poisson process of rate $\lambda_n/\bar{p}_n$, $n \geq 1$. The r.v. W_n has exponential $\exp(\alpha_n \bar{p}_n^{-1})$ distribution, $\bar{p}_n = (1 - p_n)$ and X_0 as well as U_n's are as in example 5.1. It is interesting to note that even in the modified setup, X_n has $G(\nu_0, M_n)$ distribution. The transition kernels given by (2.4) for the models described in examples 5.1 and the current example coincide. Thus, under both the models, the gamma branching sequences have the same family of finite dimensional distributions.

We have, therefore, two different stochastic mechanisms to generate a gamma branching sequence. Adke and Balkrishna (1992) have given a third mechanism to generate a similar non-branching gamma Markov sequence. All these methods create a problem of identifying the underlying stochastic structure governing the $\mathcal{X}$- processes on the basis of probability measures induced by them. In fact it is easy to establish that the probability measures induced by the $\mathcal{X}$-processes according to the three different methods are completely mutually non-identifiable in the sense of definition 2 or Puri (1985), p. 731. One way of resolving this problem of non-identifiability, at least in the case of branching gamma sequences is to follow Puri (1985) and associate the $\mathbf{N}(\mathcal{X})$-process with the $\mathcal{X}$-process. The following discussion brings out the fact that the probability measures induced by the $(\mathcal{X}, \mathbf{N}(\mathcal{X}))$-processes under the above two models are identifiable.

It is obvious that all the results pertaining to the $\mathcal{X}$-process in example 5.1 are valid for the model under discussion. In order to obtain the distribution of $N_n(X_{n-1})$, notice that since $\{A_n(t), t \geq 0\}$ is a Poisson process and X_n has $G(\nu_0, M_n)$ distribution. $A_n(X_{n-1})$

has $NB(\nu_0, \lambda_n/(M_{n-1}\bar{p}_n + \lambda_n))$ distribution, $n \geq 1$. Consequently, $N_n(X_{n-1})$ is distributed like sum of $A_n(X_{n-1})$, i.i.d. geometric $NB(1, \bar{p}_n)$ r.v.s.

The $\mathcal{X}$-process induces the sequences $\{N_n(A_n(X_{n-1})), \ n \geq 1\}$ and $\{A_n(N_n(X_{n-1})), n \geq 1)$ with state-space Z^+. Using the argument similar to that in theorem 2.1, one can show that these are also branching processes with immigration in varying environments.

The properties of the above defined BP's under the assumption A.6 can also be easily obtained. It is possible to demonstrate the existence of stationary distributions for each one of them. The stationary BP's are all distinct and new. Similarly, the probabilities of extinction by the n-th generation and of ultimate extinction, when immigration is absent can also be computed. We omit the details.

In the previous examples 5.1 and 5.2, the $\mathcal{X}$-process had continuous state-space $[0, \infty)$. The following example 5.3 deals with the situation when $\mathcal{X}$-process has Z^+ as its state-space.

Example 5.3. Let $\mathbf{W}_n$ be a sequence of i.i.d. r.v.s. W_{in}, $i \geq 1$ each with the logarithmic series distribution defined by

$$P[W_{in} = j] = (-\log \bar{p}_n)^{-1} p_n^j / j, \quad j = 1, 2, \cdots$$

where $0 < p_n < 1$, $n \geq 1$. The driving process $\mathbf{N}_n$ is assumed to be the homogeneous Poisson process of rate $(-\log \bar{p}_n)$ and the immigration r.v.s U_n have $NB(\nu_0, p_n)$ distribution, $n \geq 1$. The initial r.v. X_0 is also assumed to have $NB(\nu_0, p_0)$ distribution. The recurrence relation (2.5) for the above process is

$$L_{n+1}(s) = \left[\frac{\bar{p}_{n+1}}{1 - p_{n+1}e^{-s}}\right]^{\nu_0} L_n\left(-\log \frac{\bar{p}_{n+1}}{1 - p_{n+1}e^{-s}}\right), \ n \geq 0.$$

This relation can be used, after some routine algebra and induction argument to show that X_n has $NB(\nu_0, \pi_n^{-1})$ distribution, π_n's are defined by

$$\pi_0 = 1/p_0, \ \pi_n = 1 + (p_n/\bar{p}_n)\pi_{n-1}, \ n \geq 1$$

The r.v. $N_n(X_{n-1})$ is distributed like sum of X_{n-1} i.i.d. Poisson r.v.s, each with mean $(-\log \bar{p}_n)$.

If $p_n = p < 1/2, n \geq 1$ and $p_0 = (1 - 2p)/(1 - p)$, then the $\mathcal{X}$-process is a stationary negative binomial BP in which, each X_n has $NB(\nu_0, (1 - 2p)/(1 - p))$ distribution. The stationary distribution of $\mathbf{N}(\mathcal{X})$-process is that of the sum of a random number of i.i.d. Poisson r.v.s each with mean $(-\log \bar{p})$, the number of summands having a $NB(\nu_0, (1 - 2p)/(1 - p))$ distribution.

The model discussed by McKenzie (1986) is a particular case of our model (1.2). Thus one has two different approaches to the construction of stationary negative binomial

B.P.s. However, one does not encounter the problem of identifiability between these two models as their transition kernels are different. Nevertheless, the immigration component (innovation component in the time series parlance) specified by McKenzie seems to be contrived one in comparison with the one in the present example.

6   Concluding Remarks

The model introduced in this paper is a versatile model in the sense that by an appropriate choice of the probability laws governing the component processes, one can generate a variety of new processes with novel properties. Thus, for example, if for some fixed integer $k > 0$, $(\mathbf{N}_{n+k}, \mathbf{W}_{n+k}, U_{n+k})$ has the same distribution as that of $(\mathbf{N}_n, \mathbf{W}_n, U_n)$ for each $n \geq 1$, we get a seasonally stationary BP with k seasons. If we choose $N_n(t) \equiv 0$ a.s. whenever n is a multiple of $(m + 1)$, and is non-degenerate otherwise for $t > 0$, then one has a m-dependent BP. These types of BPs have potential application in biological growth models.

The results of this paper heavily depend on the assumptions that (a) the $\mathbf{N}_n$-processes have state-space Z^+ and (b) they have stationary and independent increments. In case the offspring and immigration r.v.s W_n and U_n, respectively, are both non-negative integer-valued, we may use Z^+ or $\mathbf{R}^+$ for the index set of the $\mathbf{N}_n$-processes. When W_n and/or U_n are not integer-valued, the index set has to be necessarily $\mathbf{R}^+$. A useful class of $\mathbf{N}_n$-processes is provided by the exponential dispersion models investigated by Jorgenson (1992). When $t \epsilon Z^+$, we do not necessarily need the infinite divisibility of the $N_n(1)$, r.v.s to preserve their s.i.i. property.

The broad branching feature of the $\mathcal{X}$-process is preserved even when $\mathbf{N}_n$-processes do not have s.i.i. property. In particular one can show that the $\mathcal{X}$- and $\underline{N}(\mathcal{X})$-processes continue to be Markov sequences without any restrictions on the $\mathbf{N}_n$-processes except that their state-space is Z^+. If the $\mathbf{N}_n$-processes are allowed to be arbitrary ones, then the $\mathbf{N}(\mathcal{X})$ process need not be a GWI process. Further work in this direction is in the progress.

REFERENCES

1. **ADKE, S. R. AND BALAKRISHINA, N.** (1992). *Markovian chi-square and Gamma processes*, Stat. Prob. Letters, 15, 349–356.

2. **AL-OSH, M. A. AND ALY, E. A. A.** (1992). *First order auto-regressive time series with negative binomial and geometric marginals*, Commun. Statist. Theory Math., 21, 2483–2492.

3. **AL-OSH, M. A.. AND ALZAID, A. A.** (1987). *First order integer valued auto-regressive (INAR(1)) process*, J. Time Series Anal., 8, 261–275.

4. **HEATHCOTE, C. R.** (1965). *The branching process allowing immigration*, J. Roy. Statist. Soc., B27, 138–173, Corr. ibid (1966), B28, 213–213.

5. **JAGERS, P.** (1975). *Branching processes with biological applications*, Wiley, London.

6. **JORGENSEN, B.** (1992). *Exponential dispersion models and extensions : a review*, Int. Statist. Review, 60, 1, 5–20.

7. **KALLENBERG, P. J. M.** (1979). *Branching processes with continuous state space*, Math. Centre Tracts 117, Amsterdam.

8. **LOÈVE, M. M.** (1977). *Probability Theory I*, Fourth Edn., Springer-Verlag.

9. **MCKENZIE, E. D.** (1986). *Auto-regressive moving average processes with negative binomial and geometric marginal distributions*, Adv. Appl. Prob., 18, 679–705.

10 **PURI, P. S.** (1986). *On some non-identifiability problems among some stochastic models in reliability theory in : Proc. Berkeley Conf. in honor of J. Neyman and J. Kiefer*, Vol. II, Ed. Le Cam, C. M. an Olshen, R. A., Wadsworth, Monterey, Calfornia, USA, 729–748.

11 **SIM, C. H.** (1990). *First order auot-regressive models for gamma and exponential processes*, J. Appl. Prob., 27, 325–332.

The extinction time of the inhomogeneous branching process

J.C. D'SOUZA[1]

Heriot-Watt University,
Edinburgh, U.K.

Abstract

Let $\{Z_n\}$ be a Galton-Watson process in varying environments. Under a condition which ensures that the process dies out almost surely we give asymptotic formulae for $P(Z_n > 0)$.

AMS 1980 subject classification: Primary 60J80
Keywords and phrases: Branching processes, Galton-Watson process, extinction time, varying environments.

1 Introduction.

We consider the Galton-Watson process in varying environments (GWPVE), that is a time-inhomogeneous Galton-Watson branching processes $\{Z_n\}$, defined by

$$Z_0 = 1, \quad Z_{n+1} = \sum_{i=1}^{Z_n} X_{n,i} \qquad (n \geq 0),$$

where $\{X_{n,i}; i\}$ are independent identically distributed copies of a random variable X_n, taking values on the non-negative integers.

We set

$$\mu_n := EX_n, \quad m_n := EZ_n,$$

and the generating functions ϕ_n, f_n and f_n^j are defined by

$$\phi_n(s) := Es^{X_n}, \quad f_n^j(s) := E(s^{Z_n} \mid Z_j = 1), \text{ and } f_n(s) := f_n^0(s).$$

We assume throughout this paper that the offspring means are finite and non-zero, so for all n

$$0 < \mu_n < \infty.$$

It is then standard that for $n \geq k$

$$\frac{m_n}{m_k} = \prod_{j=k}^{n-1} \mu_j,$$

[1]Present address: Department of Mathematical Sciences, University of Aberdeen, Aberdeen AB9 2TY, UK.

and that for all $j \le k \le n$

$$f_n^j(s) = f_k^j(f_n^k(s)).$$

We shall impose a condition to ensure that the process dies out almost surely, so that

$$P(Z_n \to 0) = 1,$$

and study the rate at which the quantity $P(Z_n > 0)$ goes to zero.

This problem has been studied in the case of the homogeneous Galton-Watson Process (GWP), where the $\{X_{n,i}\}$ are identical copies of a random variable X. The standard results are that if $\mu := EX < 1$, (the subcritical case), then $P(Z_n > 0)/\mu^n$ decreases to a limit which is positive if and only if

$$EX(\log^+ X) < \infty,$$

whereas if $\mu = 1$, (the critical case) then

$$nP(Z_n > 0) \to 2\sigma^{-2}$$

where σ^2 denotes the variance of X, and the right hand side has the obvious interpretation if σ^2 is zero or infinite. These results can be found in the books by Athreya and Ney (1972), Jagers (1975) and Asmussen and Hering (1983).

Rather less is known about the GWPVE. Fujimagari (1980) has considered the problem, making assumptions that ensure that the sequence $\{\mu_n\}$ converges to a limit and that the variances of the $\{X_n\}$ are finite and bounded. (Fujimagari's paper also deals with processes which may survive with positive probability, but we do not consider these processes here.)

The aim of this paper is to improve Fujimagari's results under weaker assumptions. We do not assume that $\{\mu_n\}$ converges, and we will allow the offspring variances to be infinite.

We will need a condition on the offspring means to ensure that the process cannot survive. The results of Section 2 are proved under the condition that there is a positive A and a $c > 1$ such that for all n and k

$$m_{n+k}/m_k \le Ac^{-n}. \tag{1.1}$$

This implies that

$$\liminf_{n\to\infty} \mu_n \le c^{-1} < 1.$$

In Section 3 we do not assume (1.1) and include the case where $\{\mu_n\}$ may converge to one. Most of these results are proved using methods which have been used to decipher the behaviour of the supercritical GWPVE. The results of Section 2 are related to the results of Goettge (1976) and D'Souza and Biggins (1992), and those of Section 3 to those of D'Souza (1994). These results are proved in the final section.

2 Exponential decay of means

In this section we assume that $\{m_n\}$ converges to zero at a rate which is at least exponential. We impose the condition that for all non-negative n and k

$$\frac{m_{n+k}}{m_k} \le \frac{A}{g(n)}, \tag{2.1}$$

where A is a positive constant and g is a differentiable function with $g(0)$ equal to one, which strictly increases to infinity. We also assume that $g^{-1}(x)$, which must exist for x greater than or equal to one, has the representation

$$g^{-1}(x) = \int_1^x \frac{L(y)}{y} \, dy, \tag{2.2}$$

for some function L slowly varying at infinity. For the theory of slowly varying functions the reader is referred to the books by Bingham et al (1987) or Seneta (1986).

It can be shown that (2.1) implies that

$$\limsup_{n \to \infty} m_n^{1/n} < 1,$$

so we may assume that

$$g(n) \geq c^n$$

for all large n and some $c < 1$, so that (1.1) holds.

Equation (2.2) includes a wide range of functions. For example, this includes the class $g(x) = \exp(x^{1/r})$, for positive r when

$$L = r(\log x)^{r-1}$$

and $g(x) = \exp\{\exp\{x^{1/r}\} - 1\}$, for positive r when

$$L(x) = \frac{r \left(\log(1 + \log x)\right)^{r-1}}{(1 + \log x)}.$$

Under (2.1) we will give a condition to ensure that

$$0 < \inf_n \left\{ P(Z_n > 0)/m_n \right\} \leq \sup_n \left\{ P(Z_n > 0)/m_n \right\} \leq 1.$$

The right hand side of this inequality is immediate, so Theorem 1 handles the left hand side. In order to prove the left hand side of this inequality we need to impose a condition on the tail behaviour of the X_n. This requires a definition. We say that a random variable X dominates a sequence of variables $\{V_n\}$ if, for all n and all positive x

$$P(X \geq x) \geq P(V_n \geq x).$$

Theorem 1 *Let $\{Z_n\}$ be a GWPVE satisfying (2.1). Suppose that the sequence $\{(X_n \mid X_n > 0)\}$ is dominated by a random variable X, satisfying $E(Xg^{-1}(X)) < \infty$. Then*

$$\inf_n \frac{P(Z_n > 0)}{m_n} > 0. \tag{2.3}$$

Notice that when we specialize this theorem to the case of the subcritical GWP we take $g(x) = \mu^{-x}$, so we are asserting that a sufficient condition for the conclusion of the theorem to be valid is that

$$EX(\log^+ X) < \infty. \tag{2.4}$$

This is the well known result stated in the introduction, and, in fact, (2.4) is also necessary for (2.3) to hold for the GWP, so Theorem 1 is quite tight.

Without a moment condition we may have that

$$\lim_{n\to\infty} P(Z_n > 0)/m_n = 0,$$

and this may converge at a very fast rate. For example, take a strictly greater than one, and μ less than one and consider the process defined by

$$\phi_n(s) = \left(1 - a^{-n}\right) + a^{-n} s^{[a^n \mu]} \quad \text{for } n > n_0 = (-\log\mu)/(\log a),$$

with $\phi_n(s)$ equal to s for $n \leq n_0$. It is clear that (1.1) holds, but

$$
\begin{aligned}
P(Z_n > 0) &= \sum_{j=1}^{\infty} P(Z_{n-1} = j)\left\{1 - \left(1 - a^{-n+1}\right)^j\right\} \\
&\leq \sum_{j=1}^{\infty} P(Z_{n-1} = j) j a^{-n+1} \\
&= a^{-n+1} E Z_{n-1},
\end{aligned}
$$

so

$$\limsup_{n\to\infty} \{P(Z_n > 0)/m_n\}^{1/n} \leq a^{-1} < 1.$$

This can not happen in the case of the subcritical GWP when we must have

$$\{P(Z_n > 0)/\mu^n\}^{1/n} \to 1,$$

and the next theorem gives a condition which ensures that the same result holds for the GWPVE.

Theorem 2 *Let* $\{Z_n\}$ *be a GWPVE satisfying (2.1). Suppose that the sequence* $\{(X_n \mid X_n \geq 1)\}$ *is dominated by a random variable* X, *satisfying* $EX < \infty$. *Then*

$$\{P(Z_n > 0)/m_n\}^{1/n} \to 1.$$

Remark

There are similarities between the subcritical and supercritical GWPVEs. For example, suppose that for all n and k

$$\frac{m_{n+k}}{m_k} \geq Ag(n),$$

where A is a positive constant and g is a function satisfying the conditions of this section. It can be shown that if $\{X_n/\mu_n\}$ is dominated by a random variable X and EX is finite, then the process is supercritical in that it survives with a positive probability. If $E(Xg^{-1}(X))$ is finite, then W, the limit of the non-negative martingale $\{Z_n/m_n\}$, has expectation equal to one, and

$$\{W > 0\} \equiv \{Z_n \to \infty\} \text{ almost surely.}$$

Furthermore, if $E(Xg^{-1}(X))$ is infinite, but EX is finite then we have the weaker conclusion that

$$\left\{(Z_n/m_n)^{1/n} \to 1\right\} \equiv \{Z_n \to \infty\} \text{ almost surely.}$$

D'Souza and Biggins (1992) proved this for the special case where g is the exponential function. (See also Goettge (1976) for related results). The extension to general g uses their method and Lemma 2 below.

3   Non-exponential decay of means

In this section we no longer assume that (1.1) holds, as we will allow for the case where μ_n may converge to one. Because of this we need a stronger moment condition on the offspring distributions than that imposed in Theorem 1. We impose the condition that there is a $p \in (1,2]$ and a finite constant K such that

$$E(X_n/\mu_n)^p < K \quad \text{for all } n \geq 0. \tag{3.1}$$

It will be convenient to write $\beta = 1/(p-1) \geq 1$.

Theorem 3 *Let $\{Z_n\}$ be a GWPVE. Then*

(i)

$$\frac{P(Z_n > 0)}{m_n} \leq \left\{ 1 + m_n \sum_{j=0}^{n-1} \frac{\mu_j - P(X_j \geq 1)}{m_{j+1}\mu_j} \right\}^{-1}.$$

(ii) Suppose that (3.1) is satisfied. Then

$$\frac{P(Z_n > 0)}{m_n} \geq \left\{ m_n^{p-1} + (K-1) \sum_{j=0}^{n-1} \left(\frac{m_n}{m_j} \right)^{p-1} \right\}^{-\beta}.$$

The second part of this theorem shows that if (3.1) is satisfied

$$P(Z_n > 0) \geq \left\{ 1 + (K-1) \sum_{j=0}^{n-1} m_j^{-(p-1)} \right\}^{-\beta},$$

so a sufficient condition for the processes to survive with positive probability is

$$\sum_{j=0}^{\infty} m_j^{-(p-1)} < \infty.$$

The special case when $p = 2$ is well known. (See Theorem 2 of Agresti (1975).)

We now give specific examples of GWPVEs, and use this theorem to find bounds for their extinction probabilities. As well as controlling the right hand tails of the offspring distributions we will need to control the left hand tails. Fujimagari (1980) did this by assuming that

$$\limsup_{n \to \infty} \frac{P(X_n \geq 1)}{\mu_n} < 1. \tag{3.2}$$

This implies that for all large j

$$\frac{\mu_j - P(X_j \geq 1)}{\mu_j} \geq \epsilon$$

where ϵ is a positive constant. Then, Theorem 3 shows that

$$P(Z_n > 0) \leq \left\{ m_n^{-1} + C + \sum_{j=0}^{n-1} \frac{\epsilon}{m_{j+1}} \right\}^{-1},$$

for some constant C, so a sufficient condition for the process to die out almost surely is

$$\sum_{j=0}^{\infty} m_{j+1}^{-1} = \infty. \tag{3.3}$$

This result is due to Fearn (1976), whilst Theorem 2 of Agresti (1975) and Theorem 2.2 of Jirina (1976) contain very similar results. In the examples given below we have, therefore, imposed conditions which ensure that the sum in (3.3) is infinite.

Example 1 *Let $\{Z_n\}$ be a GWPVE.*

(i) *Suppose that there exists $B > 0$ and $\delta > -1$ such that for all $n \geq k \geq 1$*

$$m_n/m_k \geq B \left(\frac{k}{n} \right)^{\delta}, \tag{3.4}$$

and (3.2) holds. Then

$$\limsup_{n \to \infty} \frac{nP(Z_n > 0)}{m_n} < \infty.$$

(ii) *Suppose that there exists $A > 0$ and $\gamma > -1$ such that for all $n \geq k \geq 1$*

$$m_n/m_k \leq A \left(\frac{k}{n} \right)^{\gamma} \tag{3.5}$$

and (3.1) is satisfied. Then

$$\liminf_{n \to \infty} \frac{n^{\beta} P(Z_n > 0)}{m_n} > 0.$$

Corollary 1 *Let $\{Z_n\}$ be a GWPVE which satisfies (3.1) and (3.2) such that for some $\delta \geq \gamma > -1$*

$$1 - \delta n^{-1} \leq \mu_n \leq 1 - \gamma n^{-1}$$

for all large n, Then there are positive constants a and b such that for all large n

$$an^{-\beta} \leq P(Z_n > 0)/m_n \leq bn^{-1}.$$

Notice that putting $p = 2$ into the corollary, so that $\beta = 1$, gives a result which contains Theorem 4.2(c) of Fujimagari (1980).

Example 2 *Let $\{Z_n\}$ be a GWPVE.*

(i) *Suppose that (3.4) is satisfied with $\delta = -1$, and that (3.2) holds. Then*

$$\limsup_{n \to \infty} \frac{n(\log n)P(Z_n > 0)}{m_n} < \infty.$$

(ii) Suppose that (3.5) is satisfied with $\gamma = -1$, and that (3.1) holds. Then

$$\liminf_{n \to \infty} \frac{n(\log n)P(Z_n > 0)}{m_n} > 0 \qquad \text{if } p = 2, \text{ and}$$

$$\liminf_{n \to \infty} \frac{n^\beta P(Z_n > 0)}{m_n} > 0 \qquad \text{if } 1 < p < 2.$$

Corollary 2 *Let $\{Z_n\}$ be a GWPVE which satisfies (3.1) and (3.2). Suppose that for all large n*

$$\mu_n = 1 + n^{-1},$$

so $m_n \sim Cn$ for some positive constant C. Then there exist positive constants a and b such that for all large n,

$$\begin{array}{rcccl} a(\log n)^{-1} & \leq & P(Z_n > 0) & \leq & b(\log n)^{-1} \quad \text{if } p = 2 \text{ and} \\ an^{1-\beta} & \leq & P(Z_n > 0) & \leq & b(\log n)^{-1} \quad \text{if } 1 < p < 2. \end{array}$$

This contains the result handled by Theorem 4.2(d) of Fujimagari (1980).

Example 3 *Let $\{Z_n\}$ be a GWPVE which satisfies (3.1) and (3.2) and suppose that*

$$0 < \inf_n m_n \leq \sup_n m_n < \infty.$$

Then there exist positive constants a and b such that

$$an^{-\beta} \leq P(Z_n > 0) \leq bn^{-1}.$$

This improves Theorem 4.2(b) of Fujimagari (1980).

We prove Theorem 3 in the next section. The proofs of the examples are straightforward, and are omitted.

The upper bounds in these examples can not be improved when $p = 2$. This is shown by Fujimagari's results. It is worthwhile, therefore, looking at the lower bounds. In all these examples the lower bound improves as p increases. This raises the question as to how good these bounds are. In fact, they seem to be quite tight. For example, consider the critical GWP whose generating function satisfies

$$\phi(s) = s + (1 - s)^{1+\theta} L\left(\frac{1}{1 - s}\right), \qquad 0 < \theta < 1,$$

where L is a function which is slowly varying at infinity. It can be shown (Theorem B of Bingham and Doney, 1974) that $EX^{1+\theta}$ is finite if and only if

$$\int_0^1 \frac{L(u)}{u} du < \infty, \tag{3.6}$$

and Bojanic and Seneta (1971) have shown that if L satisfies some regularity conditions then

$$P(Z_n > 0) \sim cn^{-1/\theta} L^{-1/\theta}(n^{1/\theta}),$$

where c is a positive constant. Suppose now that the integral in (3.6) is finite and put $p = 1 + \theta$, so $\beta = 1/\theta$. Example 3 above shows that

$$\liminf_{n \to \infty} n^\beta P(Z_n > 0) > 0,$$

whereas Bojanic and Seneta's (1971) result implies that $n^\beta P(Z_n > 0)$ is a slowly varying function increasing to infinity. Thus, although Example 3 is not as strong as the known result, we can not replace the n^β term by n^α for any $\alpha < \beta$.

4   Proofs of the theorems

We need some notation. For any random variable X taking values on the non-negative integers, which has a finite mean we define the increasing function r by

$$sr(s) = E(1-s)^X - 1 + sEX, \quad (0 < s \leq 1),$$

and we define r_n similarly through $\phi_n(s)$, the generating function of X_n. Then

$$
\begin{aligned}
P(Z_n > 0 \mid Z_i = 1) &= 1 - f_n^i(0) \\
&= 1 - \phi_i(f_n^{i+1}(0)) \\
&= (1 - f_n^{i+1}(0))\left(\mu_i - r_i(1 - f_n^{i+1}(0))\right) \\
&= (P(Z_n > 0 \mid Z_{i+1} = 1))(\mu_i - r_i(P(Z_n > 0 \mid Z_{i+1} = 1))),
\end{aligned}
$$

and it follows that

$$P(Z_n > 0) = \prod_{j=0}^{n-1} \{\mu_j - r_j(P(Z_n > 0 \mid Z_j = 1))\}.$$

Thus,

$$\frac{P(Z_n > 0)}{m_n} = \prod_{j=0}^{n-1} \left\{1 - \frac{1}{\mu_j} r_j(P(Z_n > 0 \mid Z_j = 1))\right\}. \tag{4.1}$$

Before we can use this equation we will need to bound the functions r_j. The following lemmas do this.

Lemma 1 *Suppose that* $\{(X_n \mid X_n \geq 1)\}$ *is dominated by an integer valued random variable* X, *satisfying* $EX < \infty$. *Then*

$$r(s) \geq r_n(s)/\mu_n.$$

Proof

The domination condition implies that for all n and all non-negative x

$$P(X \geq x) \geq P(X_n \geq x)/P(X_n \geq 1) \geq P(X_n \geq x)/\mu_n.$$

The right hand side inequality follows because, as the $\{X_n\}$ are integer valued, μ_n is no less than $P(X_n \geq 1)$. Thus,

$$
\begin{aligned}
sr(s) &= \sum_{j=1}^{\infty} \left\{(1-s)^j - 1 + sj\right\} P(X = j) \\
&= \sum_{j=1}^{\infty} \left\{(1-s)^j - 1 + sj\right\} \{P(X > j-1) - P(X > j)\} \\
&= \sum_{j=0}^{\infty} s\left\{1 - (1-s)^j\right\} P(X > j) \\
&\geq \sum_{j=0}^{\infty} s\left\{1 - (1-s)^j\right\} P(X_n > j)/\mu_n \\
&= sr_n(s)/\mu_n.
\end{aligned}
$$

Lemma 2 *Under the conditions of Theorem 1,*

$$
\int_0^{\infty} r(A/g(n))dn < \infty \iff EXg^{-1}(X) < \infty.
$$

Proof

Make the substitution $u = Ag(n)^{-1}$ to see that

$$
\int_0^{\infty} r(A/g(n))dn = A \int_0^A \frac{r(u)}{u^2 g'(g^{-1}(A/u))} du.
$$

As

$$
\begin{aligned}
1 = \frac{d}{dx} g(g^{-1}(x)) &= \{g'(g^{-1}(x))\}\{g^{-1'}(x)\} \\
&= \{g'(g^{-1}(x))\}\{L(x)/x\},
\end{aligned}
$$

the integral becomes

$$
\int_0^{\infty} r(A/g(n))dn = \int_0^A \frac{r(u)}{u} L(Au^{-1})du.
$$

The proof of the lemma now follows from Theorem B of Bingham and Doney (1974), which asserts that this is finite if and only if

$$
E\left(X\left(\int_1^X \frac{L(y)}{y} dy\right)\right) = E\left(Xg^{-1}(X)\right) < \infty.
$$

Proof of Theorem 1

Notice that the domination condition implies that for all n,

$$
EX \geq E(X_n)/P(X_n \geq 1),
$$

so

$$
\frac{P(X_n \geq 1)}{\mu_n} \geq \frac{1}{EX} = \delta > 0.
$$

Now, for $j \leq n$

$$P(Z_n > 0 \mid Z_j = 1) \leq P(X_j \geq 1),$$

and as r_j is an increasing function,

$$
\begin{aligned}
r_j(P(Z_n > 0 \mid Z_j = 1)) &\leq r_j(P(X_j \geq 1)) \\
&= \frac{1}{P(X_j \geq 1)}\left\{E(P(X_j = 0))^{X_j} - 1 + P(X_j \geq 1)\mu_j\right\} \\
&\leq \frac{1}{P(X_j \geq 1)}\{P(X_j = 0) + P(X_j = 0)P(X_j \geq 1) \\
&\quad -1 + P(X_j \geq 1)\mu_j\} \\
&= P(X_j = 0) - 1 + \mu_j.
\end{aligned}
$$

Thus,

$$1 - \frac{1}{\mu_j}r_j(P(Z_n > 0 \mid Z_j = 1)) \geq \frac{P(X_j \geq 1)}{\mu_j} \geq \delta.$$

Choose J large enough that $r(A/g(J)) < 1$. Then Lemma 1, together with the facts that

$$P(Z_n > 0 \mid Z_j = 1) \leq m_n/m_j$$

and that r_j is an increasing function, shows that (4.1) is greater than or equal to

$$\delta^J \prod_{j=0}^{n-J}(1 - r(m_n/m_j)) \geq \delta^J \prod_{j=0}^{n-J}(1 - r(A/g(n-j)^{-1})).$$

Now,

$$
\begin{aligned}
\sum_{j=0}^{n-J} r\left(Ag(n-j)^{-1}\right) &= \sum_{j=J}^{n} r\left(Ag(j)^{-1}\right) \\
&\leq \sum_{j=J}^{\infty} r\left(Ag(j)^{-1}\right) \\
&\leq \int_{x=0}^{\infty} r\left(Ag(x)^{-1}\right) dx
\end{aligned}
$$

and this is finite, by Lemma 2. Thus, the theorem is proved.

Proof of Theorem 2

The proof of this theorem follows directly from (4.1) and the fact that

$$\mu_j^{-1}r_j(m_n/m_j) \leq r(A/g(n-j)) \to 0 \text{ as } (n-j) \to \infty.$$

Proof of Theorem 3

Corollary 1 of Fearn (1976) shows that

$$1 - f_n(s) \leq \left\{(1-s)^{-1}m_n^{-1} + \sum_{j=0}^{n-1}\frac{\mu_j - 1 + P(X_j = 0)}{m_{j+1}\mu_j}\right\}^{-1}.$$

Put $s = 0$ in this equation to prove the first part.

To prove the second part let $W_n = Z_n/m_n$. An application of the lemma on Page 229 of Neveu (1987) shows that

$$E(Z_n^p \mid Z_{n-1}) \leq (Z_{n-1}\mu_{n-1})^p + Z_{n-1}(EX_{n-1}^p - \mu_{n-1}^p)$$

so if (3.1) holds we must have that for all $n \geq 0$

$$E(W_n^p) - E(W_{n-1}^p) \leq (K-1)m_{n-1}^{1-p}.$$

Thus

$$E(W_n^p) \leq \left\{ 1 + (K-1)\sum_{j=1}^{n} m_{j-1}^{1-p} \right\}.$$

Now, Hölder's Inequality implies that

$$
\begin{aligned}
1 = E(W_n) &= E(W_n I(W_n > 0)) \\
&\leq (E(W_n^p))^{1/p} P(W_n > 0)^{(p-1)/p},
\end{aligned}
$$

so,

$$
\begin{aligned}
P(Z_n > 0) &= P(W_n > 0) \\
&\geq E(W_n^p)^{-\beta} \\
&\geq \left\{ 1 + (K-1)\sum_{j=0}^{n-1} m_j^{1-p} \right\}^{-\beta}.
\end{aligned}
$$

This proves the theorem.

References

[1] Agresti, A. (1975) On the extinction time of varying and random environment branching processes. *J. Appl. Prob.* **12**, 39-46.

[2] Asmussen, S. and Hering, H. (1983) *Branching Processes.* Birkhäuser, Boston.

[3] Athreya, K. B. and Ney, P. E. (1972) *Branching Process.* Springer-Verlag, Berlin.

[4] Bingham, N. H. and Doney, R. A. (1974) Asymptotic properties of supercritical branching processes. I: The Galton-Watson Process. *Adv. Appl. Prob.* **6**, 711-731.

[5] Bingham, N. H., Goldie, C. M. and Teugels, J. L. (1987) *Regular Variation.* Cambridge University Press, Cambridge.

[6] Bojanic, R. and Seneta, E. (1971) Slowly varying functions and asymptotic relations. *J. Math. Anal. Appl.* **34**, 302-315.

[7] D'Souza, J. C. (1994) The rates of growth of the Galton-Watson process in varying environments. *Adv. Appl. Prob.* **26**, 698-714.

[8] D'Souza, J. C. and Biggins, J. D. (1992) The supercritical Galton-Watson process in varying environments. *Stoch. Proc. Appl.* **42**, 39-47.

[9] Fearn, D. H. (1976) Probability of extinction of critical generation-dependent Galton-Watson processes. *J. Appl. Prob.* **13**, 573-577.

[10] Fujimagari, T. (1980) On the extinction time distribution of a branching process in varying environments. *Adv. Appl. Prob.* **12**, 350-366.

[11] Goettge, R. T. (1976) Limit theorems for the supercritical Galton-Watson process in varying environments. *Math. Biosci.* **28**, 171-190.

[12] Jagers, P. (1975) *Branching Processes with Biological Applications.* Wiley, Chichester.

[13] Jirina, M. (1976) Extinction of non-homogeneous Galton-Watson Processes. *J. Appl. Prob.* **13**, 132-137.

[14] Neveu, J. (1987) Multiplicative martingales for spatial branching processes. In *Seminar in Stochastic Processes.* ed. E. Cinlar, K. L. Chung and R. K. Getoor. Progress in Probability and Statistics **15**, 233-241, Birkhäuser, Boston.

[15] Seneta, E. (1976) *Regularly Varying Functions.* Lecture Notes in Maths. No. 508 Springer, Berlin.

SMITH-WILKINSON BRANCHING PROCESSES WITH RANDOM ENVIRONMENT AND DEPENDENCE WITH COMPLETE CONNECTIONS

MARIUS IOSIFESCU, Romanian Academy *

Abstract

The Markov dual process of a Smith-Wilkinson branching process with random environment (SWBPRE for short) is used to illustrate an unusual phenomenon in the theory of dependence with complete connections (see [3]). It also appears that the SWBPRE is to be associated with fractals and image analysis. As a matter of fact, in this short note we develop ideas just sketched in Example 4 and Problem 9 in Chapter 1 of [3].

Key words: BRANCHING PROCESSES IN RANDOM ENVIRONMENT; DUAL PROCESSES; DEPENDENCE WITH COMPLETE CONNECTIONS; FRACTALS

1 Random composition of maps and fractal images

Let (W, d) be a metric space and denote by $C(W)$ the collection of all bounded continuous complex-valued functions on W. As is well known, $C(W)$ is a Banach space under the supremum norm $|.|$ defined as

$$|f| = \sup_{w \in W} |f(w)|, f \in C(W).$$

Let $\epsilon = (\epsilon_n)_{n>1}$ be a sequence of i.i.d. random variables taking the values $1, 2, ... r$ with probabilities $P(\epsilon_1 = i) = p_i > 0, 1 \leq i \leq r, \sum_{i=1}^{r} p_i = 1$, and let $\Phi = (\phi_i)_{1 \leq i \leq r}$ be a collection of continuous maps from W into itself. Consider the Markov process $Z^{w_0} = (z_n^{w_0})_{n \geq 0}$ with $z_0^{w_0} = w_0$ (arbitrarily given in W) and the following transition mechanism: given $z_n^{w_0}$ we have $z_{n+1}^{w_0} = \phi_{\epsilon_{n+1}}(z_n^{w_0})$, that is $z_{n+1}^{w_0} = \phi_i(z_n^{w_0})$ with probability $p_i, 1 \leq i \leq r$, for all $n \geq 0$. Therefore Z^{w_0} is a process of random composition of the maps of Φ: as a matter of fact, we have

$$z_n^{w_0} = \phi_{\epsilon_n} \circ .... \circ \phi_{\epsilon_1}(w_0), n \geq 1 \tag{1}$$

where $\circ$ denotes composition of maps. This is precisely the object of study in what is usually called 'fractal modelling of real world images'. See,e.g. [1], [6, Chapter5], and [8].

In this context, one is primarily interested in the convergence in distribution of Z^{w_0} to a limit probability measure μ on B_W, the σ-algebra of Borel subsets of W, whatever

---

*postal address: Romanian Academy, Centre of Mathematical Statistics, Bd. Magheru 22, RO-70158 Bucharest 22, Romania

$w_0 \in W$. This can be expressed very simply in terms of the transition operator U of Z^{w_0}, defined as

$$Uf(w) = E(f(z_i^{w_0})|z_1^{w_0} = w) = \sum_{i=1}^{r} p_i f(\phi_i(w)), w \in W,$$

for any measurable complex-valued function f on W. Note that U takes $C(W))$ boundedly into itself (in other words, Z^{W_0} is a Feller Process). The convergence alluded to above amounts to

$$\lim_{n \to \infty} U^n f(w) = \int_W f d\mu \qquad (2)$$

for all $w \in W$ and $f \in C(W)$, where U^n, $n \geq 1$, is the nth power of U given by the equation

$$
\begin{aligned}
U^n f(w) &= E(f(z_{n+1}^{w_0})|z_1^{w_0} = w) \\
&= \sum_{i_1,\ldots,i_n=1}^{r} p_{i_1}\ldots p_{i_n} f(\phi_{i_n} \circ \ldots \circ \phi_{i_1}(w)) \\
&= E(f(\phi_{\epsilon_n} \circ \ldots \circ \phi_{\epsilon_1}(w))), w \in W.
\end{aligned}
$$

The point is that, when $W = R^2$, by appropriately choosing Φ and the $p_i, 1 \leq i \leq r$, the support of μ can offer a wide range of fractal greytone images. Conversely, given a fractal greytone image $F \subset R^2$, one can choose Φ and the $p_i, 1 \leq i \leq r$, so that the support of the corresponding limiting μ be arbitrarily close to F. A sufficient condition implying (2) is given by the theorem below.

Theorem (Kaijser [5], Barnsley and Elton [1]). Assume that (W, d) is complete. Then (2) holds if, for some $0 \leq r \leq 1$,

$$E(\log \frac{d(z_1^{w'}, z_1^{w''})}{d(w', w'')}) \leq \log r \qquad (3)$$

for all $w', w'' \in W$.

Let us note that the framework just described, in which Z^{W_0} has been defined, is a special case of what is called 'dependence with complete connections', a topic initiated and mainly developed by Romanian probabilists. For details we refer the reader to [3, Chapter 4].

2 Random Composition of Maps and the Dual Process of an SWBPRE

It is natural to ask what happens when condition (3) above does not hold. We shall show that there are cases where $z_n^{w_0}$ converges in distribution as $n \to \infty$ to a limit (depending on w_0) for any $w_0 \in W$ under no extra assumptions.

The reader acquainted with branching process theory has certainly noticed that the process Z^{w_0} should be related to the SWBPRE. Let us recall that this is a non-negative integer valued process $X = (X_n)_{n \geq 0}$, where X_n denotes the population size at time n. Reproduction is affected by a sequence of i.i.d. environment variables $\epsilon = (\epsilon_n)_{n \geq 1}$ taking on values in some measurable space $(E, \mathcal{E})$: for any $n \geq 0$, given ϵ and $X_0, X_1, .., X_n$, if $X_0 \neq 0$, then the family sizes of the X_n individuals at time n are i.i.d. random variables each with a probability distribution which is determined by ϵ_{n+1} via a probability generating function $\phi_{\epsilon_{n+1}}$, and X_{n+1} is just the sum of those family sizes; if $X_n = 0$, then $X_{n+1} = 0$. Let $q(\epsilon)$ denote the probability, conditional on ϵ, that the population becomes extinct starting with a single individual, i.e. that $X_n = 0$ for some $n \geq 1$ with $X_0 = 1$. By the very definition of the process, the unconditional extinction probability starting with k individuals is $q_k = E(q^k(\epsilon))$, where the expectation is taken over the environment ϵ.

As is well known (see e.g. [4, pp.81-84]) for an *arbitrary* environment ϵ, X_n converges a.s. as $n \to \infty$ to a random variable with possible values $\infty, 0, 1, ...$. This implies that the probability generating function of X_n given $X_0 = k$, which equals $E((\phi_{\epsilon_1} \circ ... \circ \phi_{\epsilon_n}(s))^k)$, $s \in [0, 1)$, converges as $n \to \infty$ to a (possibly defective) probability generating function on the right-open interval $[0, 1)$, whatever $k \geq 1$. In turn, this implies that the random variable

$$\psi_n(s) = \phi_{\epsilon_1} \circ ... \circ \phi_{\epsilon_n}(s)$$

converges in distribution as $n \to \infty$ to a (proper) random variable α_s for all $s \in [0, 1)$. Since $q(\epsilon) = \lim_{n \to \infty} \psi_n(0)$ a.s. we have

$$q_k = E(q^k(\epsilon)) = E(\alpha_0^k), k \geq 1.$$

As $1 \geq \psi_n(s) \geq \psi_n(0)$ for all $n \geq 1$ and $s \in [0, 1)$, it follows that if $\alpha_0 = 1$ a.s., then $\alpha_s = 1$ a.s. for all $s \in [0, 1)$.

Now, in our special case of an i.i.d. ϵ, it is obvious that defining

$$z_n^s = \phi_{\epsilon_n} \circ ... \circ \phi_{\epsilon_1}(s) \tag{4}$$

for any $n \geq 1$ and $s \in [0, 1)$ we have

$$E(f(\psi_n(s))) = E(f(z_n^s))$$

whatever $f \in C([0, 1])$. It then follows that z_n^s converges in distribution as $n \to \infty$ to α_s whatever $s \in [0, 1)$. Clearly, for any $s \in [0, 1)$, $Z^{s_0} = (z_n^{s_0})_{n \geq 0}$, with $z_0^{s_0} = s_0$, is a $[0, 1)$-valued Markov process, the 'dual process' introduced by Smith and Wilkinson [7]. The transition probability function of Z^{s_0} does not depend on s_0 and is given by

$$P(z_{n+1}^{s_0} \in A | z_n^{s_0} = s) = P(e \in E : \phi_e(s) \in A) \tag{5}$$

for any $s \in [0.1)$ and any Borel subset A of $[0, 1)$, under an obvious measurability assumption on the probability generating functions $\phi_e, e \in E$ (which vanishes when E is at most denumerable). It is easy to see that the probability distributions of the $\alpha_s, s \in [0, 1)$, are all invariant for the transition probability function (5).

Writing $Uf(s) = E(f(z_1^{s_0}) | z_0^{s_0} = s)$, which implies $U^n f(s) = E(f(z_{n+1}^{s_0}) | z_1^{s_0} = s)$ for all $s \in [0, 1)$ and $f \in C([0, 1])$ (compare with the preceding section), on account of the well known conditions for extinction of an SWBPRE we can state the following result.

Proposition Consider an SWBPRE as described above.

(i) If $E(log\phi'_{\epsilon_1}(1))^+ < \infty$, then $E(log\phi'_{\epsilon_1}(1)) \leq 0$ implies

$$\lim_{n \to \infty} U^n f(s) = f(1), s \in [0, 1], f \in C([0, 1]),$$

while $E(log\phi'_{\epsilon_1}(1)) > 0$ and $E(-log(1 - \phi_{\epsilon_1}(0)) < \infty$ imply

$$\lim_{n\to\infty} U^n f(s) = \int_0^1 f d\mu_s, s \in [0,1), f \in C([0,1]), \tag{6}$$

for some probability measure μ_s with support contained in $[0,1)$.

(ii) If $E(log\phi'_{\epsilon_1}(1))^+ = \infty$, then $E(-log(1 - \phi_{\epsilon_1}(0))) < \infty$ implies (6).

Remarks. 1. Similar results can be stated for a multitype SWBPRE (cf. [9]).

2. Some hints of the probability measure μ_s can be found in [2] under special assumptions on E and the $\phi_e, e \in E$.

To conclude, let us note that

$$\phi'_e(1) = \sup_{s'\neq s''} \frac{|\phi_e(s') - \phi_e(s'')|}{|s' - s''|}, e \in E,$$

so that the condition $E(log\phi'_{\epsilon_1}(1)) \leq 0$ is equivalent to the condition (3) with $r = 1$. In the context of the preceding section, it appears that the iterative scheme (4) associated with an SWBPRE - which is a special case of (1) when E is a finite set - offers non-trivial limit probability measures just in case of non-extinction, that is in case of prevalently expanding maps $\phi_e, e \in E$. This circumstance is especially interesting as the assumption of prevalently it contractive maps is a standard one in the theory of dependence with complete connections.

References

[1] Barnsley, M.F. and Elton, J.H. (1988) A new class of Markov processes for image encoding. *Adv. Appl. Prob.***20**, 14-32.

[2] Grey, D.R. and Lu Zhunwei (1993) The asymptotic behaviour of extinction probability in the Smith-Wilkinson branching process. *Adv. Appl. Prob.***25**, 263-289.

[3] Iosifescu, M. and Grigorescu, S. (1990) *Dependence with Complete Connections and its Applications.* Cambridge Univ. Press, Cambridge.

[4] Jagers, P. (1975) *Branching Processes with Biological Applications.* Wiley, New York.

[5] Kaijser, T. (1978) A limit theorem for Markov chains in compact metric spaces with applications to products of random matrices. *Duke Math. J.***45**, 311-349.

[6] Peitgen, H.-O. and Saupe, D. (Eds.) (1988) *The Science of Fractal Images.* springer, New York.

[7] Smith, W.L. and Wilkinson, W.E. (1969) On branching processes in random environments. *Ann. Math. Statist.***40**, 814-827.

[8] Vrscay, E.R. (1991) Iterated function systems: theory, applications and the inverse problem. In *Fractal Geometry and Analysis*, Eds. Belair, J. and Dubuc, S., Kluwer, Dordrecht, pp. 405-469.

[9] Weissner, E.W. (1971) Multitype branching processes in random environments. *J. Appl. Prob.***8**, 17-31.

Super-Brownian Motions
in Catalytic Media

Donald A. Dawson[*]

Carleton University Ottawa

Klaus Fleischmann[†]

Institute of Applied Analysis and Stochastics, Berlin

Jean-François Le Gall[‡]

University Paris VI

ABSTRACT Some recent results on (critical continuous) super-Brownian motions X in catalytic media, i.e. where the branching rate ϱ is assumed to be a generalized function, will be reviewed. An extremely simplified one-dimensional example is $\varrho = \delta_c$. Here branching occurs only at a single point catalyst with position $c \in R$ and with infinite rate; outside c only the heat flow acts. This single point-catalytic super-Brownian motion X has remarkable properties discussed in some detail. For instance, jointly continuous super-Brownian local times $y := \{y_t(a); \, t > 0, \, a \in R\}$ exist, but $\{y_t(c); \, t > 0\} =: y(c)$ is only singularly continuous. The intuitive reason behind this is that the catalyst normally kills off the mass, by the infinite rate of branching, but "occasionally" (at exceptional times of "full" dimension) branching occurs. The super-Brownian local time $y(c)$ at c is a basic object in this model. In fact, it can be alternatively constructed as the total occupation time measure of a one-sided super-$\frac{1}{2}$-stable motion on R_+. Using $y(c)$, the mass density field $x := \{x_t(a); \, t > 0, \, a \neq c\}$ of X can then be defined by an excursion type formula, so that it solves the heat equation and is C^∞. Another problem is the construction of higher dimensional catalytic super-Brownian motions with absolutely continuous states (in contrast to the constant medium case). Some nonlinear reaction diffusion equations in which δ-functions enter in various ways are a main analytical tool.

Key words: point-catalytic medium, super-Brownian local time, sample path smoothness, total extinction, compact support, higher-dimensional absolutely continuous states

1   Model

Superprocesses are certain *measure-valued branching processes*. In principle they can be thought of as *high density approximations* to specific branching particle systems in some space.

---

[*]Department of Mathematics and Statistics, Carleton University, Ottawa, Canada K1S 5B6

[†]Institute of Applied Analysis and Stochastics, Mohrenstr. 39, D–10117 Berlin, Germany

[‡]Laboratoire de Probabilités, Université Paris VI, 4 Place Jussieu, Tour 56, F–75252 Paris Cedex 05, France

1.1 FELLER'S BRANCHING DIFFUSION

To be more precise, let us start with the simplest case where the space consists of a *single* point, so that the positions of the particles are not distinguishable and their motions can be completely neglected.

Let $z^{(\varrho)} = \{z_t^{(\varrho)}; t \geq 0\}$ denote a continuous-time *critical binary Bienaymé-Galton-Watson (BGW) process* with *branching rate* $\varrho > 0$. That is, start at time $t = 0$ with $z_0 > 0$ initial particles, and let them evolve independently in the following way: Each particle dies with rate ϱ and splits into 2 particles with rate ϱ:

$$1 \longrightarrow \begin{cases} 0 & \text{with rate } \varrho, \\ 2 & \text{with rate } \varrho. \end{cases}$$

Of course, in this extremely simplified situation this is just a *birth and death process,* and a path looks as shown in Figure 1. Now let N be a large number, start with $z_0 = [N\zeta_0]$

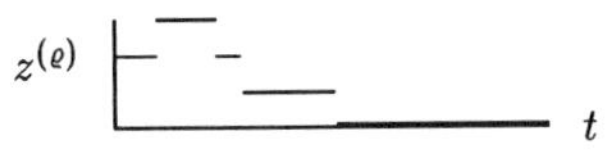

FIGURE 1. Critical binary BGW path

initial particles (where $\zeta_0 > 0$ is fixed), give each particle the small mass $1/N$, and speed up the process by switching to the large branching rate $N\varrho$. Then,

$$\left\{ \tfrac{1}{N} z_t^{(N\varrho)} \,\Big|\, z_0 = [N\zeta_0] \right\}_{t \geq 0} =: \left\{ \zeta_t^{(N)} \right\}_{t \geq 0}$$

defines a sequence $\zeta^{(N)}$ of continuous-time processes with values in R_+ which now typically look as in Figure 2. Passing to the limit in distribution as $N \to \infty$ we get the well-known

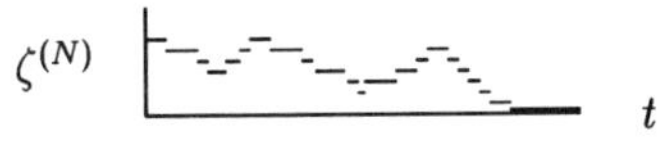

FIGURE 2. Scaled critical BGW path

(critical) branching diffusion of Feller:

$$\zeta^{(N)} \underset{N \to \infty}{\Longrightarrow} \zeta$$

[Feller (1951)]. This is a diffusion process on R_+ which can also be obtained as a solution of the stochastic equation

$$d\zeta_t = \sqrt{2\varrho\zeta_t}\, dW_t, \qquad t > 0, \; \zeta_0 \geq 0,$$

with W a one-dimensional standard Wiener process. A sample path has the typical form as shown in Figure 3.

As we have said superprocesses are certain *measure-valued* processes, so we now introduce

$$X_t := \zeta_t \, \delta_0, \qquad t \geq 0,$$

FIGURE 3. Feller's critical branching diffusion

i.e. we adjoin the population mass ζ_t to the location 0, the single point of the space considered. (Incidentally, we do not distinguish in notation between δ-measures and their generalized derivatives, the δ-functions.) This process could now be viewed as a (continuous critical) superprocess in *zero* dimensions, and we write symbolically

$$\mathrm{d}X_t = \sqrt{2\varrho X_t}\,\mathrm{d}W_t, \qquad t > 0,\ X_0 \geq 0.$$

1.2 SUPER-BROWNIAN MOTION

Going back to critical binary BGW processes, we now let the particles additionally undergo *independent standard Brownian motions* in Euclidean space R^D, where newborn particles start at their parents' position, see Figure 4. This is the so-called *critical binary branching*

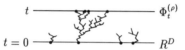

FIGURE 4. Critical binary branching Brownian motion

Brownian motion in R^D with branching rate ϱ. A useful description of the states $\Phi_t^{(\varrho)}$ at time t is given by means of counting measures:

$$\Phi_t^{(\varrho)} := \sum_{i=1}^{z_t^{(\varrho)}} \delta_{w_t^i}, \qquad t \geq 0.$$

Here $z_t^{(\varrho)}$ is the number of particles at time t which form the BGW process, and w_t^i is the position of the ith particle at time t, which arises from some Brownian path (where these paths are *not* independent).

Now consider the same *diffusion approximation*

$$X_t^{(N)} := \left\{ \tfrac{1}{N} \Phi_t^{(N\varrho)} \,\middle|\, \Phi_0 = N\delta_x \right\} \underset{N\to\infty}{\Longrightarrow} X_t,$$

i.e. start with N initial particles situated at x, branch with the large rate $N\varrho$ and give each particle the mass $\tfrac{1}{N}$. This leads to a limiting process X which is the so-called (critical continuous) *super-Brownian motion* with branching rate ϱ [Watanabe(1968)]. Consequently, the population at time t is described by a measure X_t which can also be interpreted as a *"cloud of mass"* (which in general is difficult to draw). Heuristically, X can be thought of as the solution of the *symbolic* stochastic equation

$$\mathrm{d}X_t = \frac{1}{2}\Delta X_t\,\mathrm{d}t + \sqrt{2\varrho X_t}\,\mathrm{d}W_t, \qquad t > 0,\ X_0 = \delta_x,\ x \in R^D, \qquad (1)$$

where Δ is the Laplacian and $\dot{W}$ is now a *space-time white noise*. This symbolic equation clearly exhibits the *2 components* in the model. Independently at each space point the population grows according to Feller's branching diffusion, and additionally the population mass is smeared out by the *heat flow*.

Very many properties of super-Brownian motion are already known and we mention only one: If the dimension D equals one then with probability one X lives on the space of *absolutely continuous measures*,

$$X_t(\mathrm{d}a) = x_t(a)\,\mathrm{d}a, \qquad t > 0,$$

where the *density field* x solves the stochastic equation

$$\mathrm{d}x_t(a) = \frac{1}{2}\Delta x_t(a)\,\mathrm{d}t + \sqrt{2\varrho x_t(a)}\,\mathrm{d}W_t(a), \qquad t > 0,\ a \in R, \tag{2}$$

(the one-dimensional Laplacian operator Δ acts on the space variable a). Hence in the one-dimensional case, (1) can be given a rigorous meaning. This result is due to Konno and Shiga (1988) and Reimers (1989).

Summarizing, super-Brownian motions (or more generally superprocesses) are only a particular type of spatial branching models: a continuous analogue, for instance, to branching random walks. [For a recent survey on superprocesses we refer to Dawson (1993).]

1.3 CATALYTIC MEDIA

In the form in which we introduced the super-Brownian motion, a constant enters into the model: the branching rate ϱ. Now imagine that ϱ in (1) is not a constant but varies in space: $\varrho = \varrho(a)$, $a \in R^D$. For instance in Figure 5, at some places the branching occurs

FIGURE 5. Varying branching rate

with a larger rate than at others. Such a situation is usually called a *model in a varying medium*. In this case it is not necessary that ϱ be such a smooth function but could be only a *measurable* function. Actually, ϱ may even be a *generalized function* (*"irregular" medium*). A very particular situation is given if ϱ equals a Dirac δ-function δ_c, see Figure 6. That is, branching occurs only at the single point c and there with an infinite rate. In

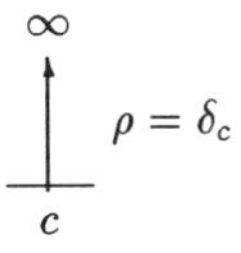

$$\infty$$
$$\big\uparrow \quad \rho = \delta_c$$
$$\underline{\hspace{2cm}}$$
$$c$$

FIGURE 6. Single point catalyst

this case we say, that a *point catalyst* is located at c which controls the branching, whereas outside c only the heat flow acts.

We can imagine $\varrho = \delta_c$ approximately as shown in Figure 7. At the particle level this means that a particle may branch only if it is within this ε-vicinity of c, see Figure 8.

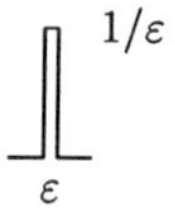

$$\prod \ ^{1/\varepsilon}$$

$$\varepsilon$$

FIGURE 7. Approximation of a δ-function

Then $\int_0^t ds\, \mathbf{1}\{|w_s - c| \leq \frac{\varepsilon}{2}\}$ represents the occupation time by time t. During that period

$$\varepsilon$$

FIGURE 8. A particle's path to the catalyst

the particle will branch with rate $\frac{1}{\varepsilon}$. Therefore the resulting quantity is

$$\frac{1}{\varepsilon}\int_0^t ds\, \mathbf{1}\left\{|w_s - c| \leq \frac{\varepsilon}{2}\right\} \xrightarrow[\varepsilon \to 0]{} L^c(t)$$

where L^c denotes the *Brownian local time* at c. But Brownian local times at points make proper sense only in dimension *one*. This suggests that point catalysts are reasonable only in one dimension.

On the other hand, there are many other generalized functions ϱ than the δ-functions, and in higher dimensions we can think of some *fractal catalytic media* ϱ, for instance. [For a discussion of fractal catalysts from a physical point of view we refer to Sapoval (1991).]

Our main reason to study branching models in *varying* media, in particular in irregular media, is to search for new phenomena caused by the medium. A very simple example of a new effect is the following: The *total mass process* $\{X_t(R^D);\ t \geq 0\}$ of a super-Brownian motion X is a Feller branching diffusion if and only if the medium is constant, that is if $\varrho = const$.

[Superprocesses in irregular media were introduced in [3] and [8].]

2 Single Point-Catalytic Super-Brownian Motion

Many very interesting phenomena already exist in the extremely simplified situation of a one-dimensional super-Brownian motion X with a single point catalyst $\varrho = \delta_c$ as mentioned above.

2.1 BASIC PROPERTIES

More precisely, we consider the process $[X, \mathbf{P}_\mu, \mu \in \mathcal{M}_f]$ in the point-catalytic medium $\varrho = \delta_c$ with law $\mathbf{P}_\mu$, where the catalyst's position $c \in R$ is fixed once and for all and μ is the initial state X_0 of X in the set $\mathcal{M}_f = \mathcal{M}_f(R)$ of finite measures on R. For the existence of this finite measure-valued Markov process we rely at this point on our intuition based on the existence of Brownian local times in one dimension. [For a strong construction of X by regularization of δ_c, see [3]; a different approach to X using the Brownian local time at c is possible via Dynkin's additive functional approach, see Dynkin (1991)].

The first result we would like to exhibit is the following theorem (see [5]).

Theorem 1 (density field) *There is a version of X (with $\varrho = \delta_c$, $c \in R$) living in the set of all absolutely continuous measures,*

$$X_t(\mathrm{d}a) = x_t(a)\,\mathrm{d}a, \qquad \text{for all } t > 0,\ \mathbf{P}_\mu\text{-a.s.},\ \mu \in \mathcal{M}_f, \tag{3}$$

where the density field $x := \{x_t(a);\ t > 0,\ a \neq c\}$ is jointly continuous. Moreover, for fixed $t > 0$ and $\mu \in \mathcal{M}_f$,

$$x_t(a) \xrightarrow[a \to c]{} 0 \quad \text{in } \mathbf{P}_\mu\text{-probability}, \tag{4}$$

whereas the variance blows up:

$$\mathbf{Var}_\mu\, x_t(a) \xrightarrow[a \to c]{} \infty, \qquad \mu \neq 0. \tag{5}$$

Consequently, the behavior of the point-catalytic model is similar to that of the model in a constant medium, except for some strange behavior at the catalyst's position: The density degenerates stochastically to 0 approaching the catalyst, in particular, the probability for the density to be large in the vicinity of the catalyst becomes very small. However, on the other hand the variance blows up approaching the catalyst.

Heuristically this can be explained as follows. The mass which will be transported into c by the heat flow will immediately be killed with "overwhelming" probability by the infinite branching rate, except on an exceptional set depending on t (where the t are uncountable). On the other hand, mass will be born only "occasionally", again by the infinite rate of branching. Both together should imply that the set of times where population mass is present at the catalyst should be *very thin* in some sense.

2.2 Super-Brownian Local Time

To put this ideas on a firm base, we introduce the so-called *super-Brownian local time* (occupation density field)

$$y_t(a) := \int_0^t \mathrm{d}s\, x_s(a), \qquad t > 0,\ a \neq c, \tag{6}$$

related to X and its (weighted) *occupation time process* $Y_t := \int_0^t \mathrm{d}s\, X_s$ (recall that $x = \{x_t(a);\ t > 0,\ a \neq c\}$ is jointly continuous by Theorem 1). Despite the "stochastic disappearance" $x_t(c) := 0$ at a fixed time t according to (4), the super-Brownian local time y remains non-degenerate as c is approached (see [5]).

Theorem 2 (non-degenerate super-Brownian local time at c) *A version of X exists such that the super-Brownian local time y defined in (6) extends continuously to all of $R_+ \times R$. The following moment formulas hold:*

$$\mathbf{E}_\mu\, y_t(a) = \int \mu(\mathrm{d}b) \int_0^t \mathrm{d}r\, p(r, a - b),$$

$$\mathbf{Var}_\mu\, y_t(a) = 2 \int \mu(\mathrm{d}b) \int_0^t \mathrm{d}r\, p(r, c - b) \left[\int_r^t \mathrm{d}s\, p(s - r, a - c) \right]^2 < \infty,$$

$\mu \in \mathcal{M}_f,\ t \geq 0,\ a \in R.$

Here p denotes the *Brownian transition density*

$$p(s, b) = \frac{1}{\sqrt{2\pi s}} \exp\left[-\frac{b^2}{2s}\right], \qquad s > 0,\ b \in R. \tag{7}$$

Hence in contrast to (5), the variance of y is always finite, whereas expectation and variance are strictly positive as long as $\mu \neq 0$. Consequently, at c a production of mass really occurs, and the super-Brownian local time y is rather "tame" compared with the x density field due to the additional integration. Now we come back to the strange behavior at c mentioned earlier.

2.3 SINGULARITY AT THE CATALYST

The super-Brownian local time y is monotonic increasing in the time variable t, and hence at each fixed position a it defines a measure λ^a on R_+:

$$\lambda^a(dt) := dy_t(a), \qquad a \in R.$$

As long as $a \neq c$, these super-Brownian local time measures λ^a are absolutely continuous, by their definition (6). But at c they behave quite differently (see [7],[5]).

Theorem 3 (singularity at c) *Let $X_0 = \delta_c$. Then with probability one the super-Brownian local time measure λ^c at c is singular and has carrying Hausdorff-Besicovich dimension one.*

Consequently, the super-Brownian local time is *singular continuous at c* (recall that this is in a sharp contrast to the constant medium case (2)). But nevertheless production of population mass occurs on a time set of *"full dimension"*.

2.4 A REPRESENTATION FORMULA FOR THE MASS DENSITY FIELD

The super-Brownian local time measure λ^c not only has very interesting properties as expressed in Theorem 3, it is actually the *basic object* in the model. For simplicity we again assume $X_0 = \delta_c$ although the results can easily be extented to general $X_0 \in \mathcal{M}_f$; see [14].

Theorem 4 (representation of x) *With $\mathbf{P}_{\delta_c}$-probability one the mass density field x can be represented as*

$$x_t(a) = \int_0^t \lambda^c(ds)\, q\big(t - s, |a - c|\big), \qquad t > 0,\ a \neq c, \tag{8}$$

where q is the deterministic function

$$q(r, b) := \frac{b}{\sqrt{2\pi r^3}} \exp\left[-\frac{b^2}{2r}\right], \qquad r, b > 0. \tag{9}$$

That is, the whole model is determined by λ^c, and moreover λ^c contains all the randomness of the model. Heuristically this can be explained as follows: According to our interpretation, mass is present at time s at the catalyst's position c according to $\lambda^c(ds)$.

Then, thinking in terms of an approximating particle system, each particle moves away from the catalyst according to a *Brownian excursion* since outside of c no branching is possible. This excursion gives a contribution to the density of mass $x_t(a)$ at time t in a if the particle is not yet back to the catalyst by time t, and moreover has exactly the position a at time t, see Figure 9. In fact, q entering in (8) is the *Brownian excursion*

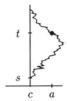

FIGURE 9. Brownian excursion away from the catalyst

density

$$q(r, b)\, \mathrm{d}b = \mathbf{n}\big\{\ell(e) > r,\ e(r) \in \mathrm{d}b\big\}, \qquad r > 0.$$

Here $\mathbf{n}$ is *Itô's excursion measure* of Brownian motion, $\ell(e)$ the length of the excursion e, and $e(r)$ its position at time r.

The representation formula (8) is a very powerful tool and has interesting applications. First of all, it implies that the total mass process dies out stochastically (see [14]):

Corollary 5 (total mass extinction) *In* $\mathbf{P}_{\delta_c}$*-probability,*

$$X_t(R) = \int_0^t \lambda^c(\mathrm{d}s)\, \sqrt{\tfrac{2}{\pi\,(t-s)}} \xrightarrow[t \to \infty]{} 0.$$

That is, although $X_t(R) > 0$, $t \geq 0$, a.s., the probability that some total mass survives as $t \to \infty$ becomes very small.

Smoothness properties of the mass density field x are another consequence of the representation formula (8) (see [14]).

Corollary 6 (smoothness of x) *There is a version of X such that with $\mathbf{P}_{\delta_c}$-probability one the mass density field x satisfies the heat equation on $\{t > 0\} \times \{a \neq c\}$, and is a C^∞-function there.*

This means that the super-Brownian motion in the point-catalytic medium is much smoother outside the catalyst than the usual super-Brownian motion (which is commonly believed to have a critical Hölder index $\frac{1}{2}$ concerning regularity in the space variable; see Perkins (1991) for a one-sided estimate). This in particular answers a question posed in [1], p. 14 in connection with some computer pictures of superprocesses. The stronger smoothness is understandable in that, according to (1) with $\varrho = \delta_c$, only the heat flow acts outside the catalyst. On the other hand, this is interesting in that we have a solution of the heat equation with a rather *irregular "boundary" condition* at $\{a = c\}$ given by the generalized derivative $\lambda^c(\mathrm{d}t)/\mathrm{d}t$ (recall that λ^c is a singular measure).

2.5 ALTERNATIVE APPROACH

Since the super-Brownian local time measure λ^c completely determines the process X, the question arises whether an alternative characterization of λ^c can be provided.

To this end, let $U = \{U_t; \, t \geq 0\}$ denote the (continuous critical) *one-sided super-$\frac{1}{2}$-stable motion* on R_+ starting with a unit mass at the origin: $U_0 = \delta_0$. That is, in the usual super-Brownian motion X as symbolically expressed in (1) with constant rate $\varrho = 1$, we replace the heat flow determined by the Brownian density p of (7) (related to $\frac{1}{2}\Delta$) by the one-sided-$\frac{1}{2}$-stable flow with transition density r,

$$r(s, b) := q(b, s), \qquad s, b > 0,$$

where q was defined in (9). (Note the interchange of the role of the time and space variables in q. Recall also that the one-sided $\frac{1}{2}$-stable process is the inverse of the Brownian local time.)

Switching to the *total occupation time* $V_\infty := \int_0^\infty dt \, U_t$ related to U we can now state the following perhaps surprising result (see [14]).

Theorem 7 (representation of λ^c) *The finite random measure λ^c coincides in distribution with V_∞. In particular, R_+ is the closed support of λ^c.*

Consequently, the super-Brownian local time measure λ^c for the point-catalytic super-Brownian motion X can be generated in law by the total occupation time measure V_∞ of the one-sided super-$\frac{1}{2}$-stable motion U on R_+. In a sense, this can be thought of as a *superprocess analog* of the classical fact that the local time measure at 0 of standard linear Brownian motion has the same distribution as the total occupation measure of a $\frac{1}{2}$-stable subordinator.

Note that via the formulas (8) and (3) first x and finally X can be defined, yielding an alternative approach to the point-catalytic super-Brownian motion:

$$U \longmapsto V_\infty \longmapsto \lambda^c \longmapsto x \longmapsto X.$$

Finally we mention that the *singularity* of the super-Brownian local time at c can be proved alternatively using this new approach to the single point-catalytic super-Brownian motion. [For details we refer to [14].]

3 Other Aspects

3.1 CUMULANT EQUATION

Remarks on some basic methods are in order. An essential tool is the so-called cumulant equation. In the case of the point-catalytic super-Brownian motion X, for instance, the *Laplace transform* of the random vector $[x_t(b), \lambda^c]$ in $R \times \mathcal{M}_f(R_+)$ can be expressed by means of solutions of the cumulant equation. To be more precise, we formulate the following proposition.

Proposition 8 (cumulant equation) *For fixed $\mu \in \mathcal{M}_f$, $\theta \geq 0$, $t > 0$, $b \neq c$, and bounded measurable non-negative function g on $[0, t]$,*

$$\mathbf{E}_\mu \exp\left[-\theta x_t(b) - \int_0^t \lambda^c(ds) \, g(s)\right] = \exp\left[-\int \mu(da) \, u(0, t, a)\right],$$

where the measurable function $u(\cdot, t, \cdot)$ is the unique solution of the integral equation (cumulant equation)

$$u(s,t,a) = \theta p(t-s, a-b) + \int_s^t dr\, p(r-s, a-c)\, g(r) - \int_s^t dr\, p(r-s, a-c)\, u(r,t,c),$$

$0 \le s < t,\ a \in R,$ *which is related to the formal "backward equation"*

$$-\frac{\partial}{\partial s} u(s,t,a) = \frac{1}{2}\Delta u(s,t,a) + \delta_c(a)\, g(s) - \delta_c(a)\, u^2(s,t,c), \qquad (10)$$

with "terminal condition" $u(s,t,\cdot)\big|_{s=t-} = \theta \delta_b$.

Note that δ-functions enter into this cumulant equation (10) at three places: as coefficient δ_c of the non-linear term describing the catalyst, into the force term $\delta_c g$ related to λ^c, and into the terminal condition related to the mass density $x_t(b)$ and which leads to fundamental solutions of (10) if $\theta = 1$. Recall that the restriction to dimension $D = 1$ is essential here.

Solutions of this equation can be constructed by means of regularization of these δ-functions; see [4].

[For a discussion of reaction diffusion equations in catalytic media from a physicochemical and biological point of view we refer to § 8.8 of Nicolis and Prigogine (1977).]

3.2 ABSOLUTELY CONTINUOUS STATES IN HIGHER DIMENSIONS

As already indicated, super-Brownian motions in two and more dimensional irregular media ϱ are a more delicate subject, since the medium may *not be arbitrarily irregular.* For instance we have to exclude a point-catalytic medium $\varrho = \delta_c$ since thinking in terms of approximating particle systems, a point catalyst will not be met by Brownian particles. On the other hand, if a not too irregular medium ϱ is *"sufficiently thin"*, then the heat flow can more effectively smear out the population mass possibly resulting in *absolutely continuous measure states* in some higher dimensions. This of course is in sharp contrast to the constant medium case, since (continuous) super-Brownian motions in dimensions $D \ge 2$ are known to have *singular* states [Dawson and Hochberg (1979)].

To demonstrate this effect, we now mention only a class of *examples.* Assume $D \ge 2$ and write the dimension D as $D = d + 1$. Moreover, let ϱ factorize to $\varrho = \varrho_d \times \varrho_1$ where ϱ_d is a random bounded measurable function on R^d (for instance a constant) and ϱ_1 is a linear combination of infinitely many δ-functions on R:

$$\varrho_1 = \sum_i \sigma_i\, \delta_{b(i)}.$$

Here we assume that the *action weights* σ_i are non-negative *i.i.d.* random variables with finite expectation, and $\sum_i \delta_{b(i)}$ is the *generalized derivative* of a *Poisson process* on R (that is a stationary Poisson point process on R). In other words, the branching rate ϱ is sampled from some random objects (branching model in a *random medium*) which can be interpreted as an *infinite collection of weighted hyperplanes* in R^D, see Figure 10.

Given a realization of the random medium ϱ one can construct a super-Brownian motion X with such irregular branching rate ϱ. That is, branching is allowed only on each hyperplane $\big\{a = [a_d, a_1] \in R^D;\ a_1 = b(i)\big\}$ (line in the figure 10), and there with unbounded rate

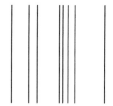

FIGURE 10. Collection of (weighted) lines in the case $D = 2$

and additional action weight σ_i. Intuitively, the existence of X is related to the existence of the non-degenerate Brownian local time (occupation density, intersection local time) at ϱ. The fact we emphasize here is that this higher dimensional catalytic super-Brownian motion X has *absolutely continuous measure states*.

We may modify this example a bit by selecting ϱ_1 as the generalized derivative of a *one-sided stable process* Γ on R with *Lévy measure* const $r^{-1-\gamma}\,dr$, where, by assuming $0 < \gamma < (2D - 1)^{-1}$, the index γ of Γ is not too close to 1. That is, we are some distance away from the constant medium case; see [13]. Then the weighted hyperplanes are even *densely situated* in R^D. Nevertheless, the states of X remain absolutely continuous. (For more general results, see [6].)

3.3 SUPPORT PROPERTIES

Finally we briefly consider the (closed) *supports* of the states of super-Brownian motions X in catalytic media. In the constant medium case it is known that if the initial measure X_0 has compact support then so too does X_t whatever the dimension is [Iscoe (1988)]. On the other hand, in the point-catalytic super-Brownian motion case of Section 2, the support of X_t, $t > 0$, is the whole space R if $X_0 \neq 0$, and the compact support property is obviously violated. The question arises whether one can formulate criteria for the compact support property to hold for super-Brownian motions in catalytic media.

In [10] one-dimensional super-Brownian motions are studied where the branching rate ϱ is selected from the generalized derivative of a *Lévy process* Λ. *Sufficient conditions* are given that the compact support property holds or is violated, respectively. *For instance*, if Λ is the one-sided stable process Γ from the previous subsection, now with general index $0 < \gamma < 1$, then the compact support property is shown to hold. On the other hand, if the Lévy measure of Λ is given by *const* $r^{-1}\mathbf{1}\{0 < r \leq 1\}\,dr$ or by a finite measure, i.e. if the δ-functions in ϱ are "too rarefied" in space, then the compact support property is lost.

CONCLUDING REMARK

Super-Brownian motions, (superprocesses; spatial branching processes) in catalytic or irregular media exhibit many interesting new effects compared with the constant medium case, and seem to merit a deeper study as indicated by the results exhibited above.

4 REFERENCES

[1] ADLER, R.J. (1994). Superprocesses: The particle picture. In: D.A. Dawson (ed.) *Measure-Valued Processes, Stochastic Partial Differential Equations, and Interacting Systems. CRM Proceedings & Lecture Notes* **5** 1-15. AMS, Providence, USA.

[2] DAWSON, D.A. (1993). Measure-valued Markov Processes. In: Hennequin, P.L. (ed.) *École d'Été de Probabilités de Saint Flour XXI-1991. Lecture Notes in Math.* **1541**, 1-260.

[3] DAWSON, D.A. and FLEISCHMANN, K. (1991). Critical branching in a highly fluctuating random medium. *Probab. Th. Rel. Fields* **90** 241-274.

[4] DAWSON, D.A. and FLEISCHMANN, K. (1992). Diffusion and reaction caused by point catalysts. *SIAM J. Appl. Math.* **52** 163-180.

[5] DAWSON, D.A. and FLEISCHMANN, K. (1994). A Super-Brownian motion with a single point catalyst. *Stochastic Process. Appl.* **49** 3-40.

[6] DAWSON, D.A. and FLEISCHMANN, K. (1994). Super-Brownian motions in higher dimensions with absolutely continuous measure states. *J. Theoretical Probab.* (to appear).

[7] DAWSON, D.A., FLEISCHMANN, K., LI, Y. and MUELLER, C. (1994). Singularity of super-Brownian local time at a point catalyst. *Ann. Probab.* (to appear).

[8] DAWSON, D.A., FLEISCHMANN, K. and ROELLY, S. (1991). Absolute continuity for the measure states in a branching model with catalysts. In: *Stochastic Processes, Proc. Semin., Vancouver, Canada 1990, Prog. Probab.* **24** 117-160.

[9] DAWSON, D.A. and HOCHBERG, K.J. (1979). The carrying dimension of a stochastic measure diffusion. *Ann. Probab.* **7** 693-703.

[10] DAWSON, D.A., LI, Y. and MUELLER, C. (1993). The support of measure valued branching processes in a random environment. Preprint, Lab. Probab. Stat., Carleton University, Ottawa.

[11] DYNKIN, E.B. (1991). Branching particle systems and superprocesses. *Ann. Probab.* **19** 1157-1194.

[12] FELLER, W. (1951). Diffusion processes in genetics. *Proc. Second Berkeley Symp. Math. Statist. Prob.* 227-246. Univ. of California Press Berkeley, California.

[13] FLEISCHMANN, K. (1994). Superprocesses in catalytic media. In: D.A. Dawson (ed.) *Measure-Valued Processes, Stochastic Partial Differential Equations, and Interacting Systems. CRM Proceedings & Lecture Notes* **5** 99-110. AMS, Providence, USA.

[14] FLEISCHMANN, K. and LE GALL, J.-F. (1994). A new approach to the single point-catalytic super-Brownian motion. Preprint No. 81, IAAS, Berlin; *Probab. Th. Rel. Fields* (submitted).

[15] ISCOE, I. (1988). On the supports of measure-valued critical branching Brownian motion. *Probab. Th. Rel. Fields* **16** 200-221.

[16] KONNO, N. and SHIGA, T. (1988). Stochastic partial differential equations for some measure-valued diffusions. *Probab. Th. Rel. Fields* **79** 201-225.

[17] NICOLIS, G., AND PRIGOGINE, I. (1977). *Self-organization in Nonequilibrium Systems. From Dissipative Structures to Order through Fluctuations.* Wiley New York.

[18] PERKINS, E. A. (1991). On the continuity of measure-valued processes. In: *Stochastic Processes, Proc. Semin., Vancouver, Canada 1990, Prog. Probab.* **24** 261-268.

[19] REIMERS, M. (1989). One dimensional stochastic partial differential equations and the branching measure diffusion. *Probab. Th. Rel. Fields* **81** 319-340.

[20] SAPOVAL, B. (1991). Fractal electrodes, fractal membranes, and fractal catalysts. In: A. Bunde and S. Havlin, eds., *Fractal and Disordered Systems* pp. 207-226. Springer, Berlin.

[21] WATANABE, S. (1968). A limit theorem of branching processes and continuous state branching processes. *J. Math. Kyoto Univ.* **8** 141-176.

MULTITYPE BRANCHING PARTICLE SYSTEMS
AND HIGH DENSITY LIMITS

LUIS G. GOROSTIZA

Centro de Investigación y de Estudios Avanzados, México

Abstract

We review some results on: long–time behavior of multitype branching particle systems and related systems of non–linear partial differential equations in the critical case, the high–density limit of such particle systems known as (multitype) measure branching process, Dawson–Watanabe process, or superprocess, and the demographic variation process.

Key words: Multitype branching particle system, superprocess, non–linear partial differential equation.

1 Introduction

We consider a class of multitype branching particle systems. Background on multitype branching processes is found in the books of Asmussen and Hering (1983), Athreya and Ney (1972), Jagers (1975), Mode (1971), and Sevastyanov (1971). Multitype systems can be regarded as monotype systems with appropriate state spaces, but the multitype framework is more suitable for our purpose.

Some recent results are concerned with the high–density asymptotic behavior of a class of continuous–time Markovian multitype branching particle systems. These results are of two sorts. In the first one there is a law of large numbers (hydrodynamic limit) and a fluctuation limit, and the main objective is to describe the fluctuation limit process, which is typically a generalized (distribution–valued) Ornstein–Uhlenbeck process. Such results are contained in López–Mimbela (1992). In the second one there is a single (non–deterministic) measure–valued limit process called measure branching process, Dawson–Watanabe process, or superprocess. The general theory of this process has attracted a great deal of attention in recent years. The lecture notes of Dawson (1992, 1993) provide an excellent introduction to the subject. We will state briefly some results for the multitype case. Multitype measure branching processes fit into the general superprocesses defined by Dynkin (1991) (see also Li, 1992), but some specific questions regarding the multitype case are not treated in the general theory. We

Postal Address: Departamento de Matemáticas, CINVESTAV, A. P. 14–740, 07000 México D. F., México.

will mention an application of critical multitype branching particle systems to obtain results on the long time behavior of solutions of a class of systems of non–linear partial differential equations. The survey paper of Wakolbinger (1994) also deals with some of these and related questions. We will end with a quick mention of the demographic variation process.

The notation we use is detailed in the Appendix.

2 The multitype branching particle system

The branching particle system on $\mathbb{R}^d$ which we consider evolves in the following way. The particles are of types $i = 1, \ldots, k$. At time $t = 0$ they are distributed in some random way. As time elapses, particles of type i migrate according to a spherically symmetric stable process with index $\alpha_i \in (0, 2]$, die at rate V_i, and at death time they produce, at death site, particles of type j chosen with probability m_{ij}, the number of which is governed by the generating function

$$\mathcal{F}_{ij}(s) = s + b_{ij}(s - 1) + c_{ij}(1 - s)^{1+\beta_{ij}}, \quad s \in [0, 1],$$

with $b_{ij} \in (-1, c_{ij}]$, $c_{ij} \in (0, (1 + b_{ij})/(1 + \beta_{ij})]$, $\beta_{ij} \in (0, 1]$. The migrations, lifetimes and branchings of different particles are independent of each other and of the initial configuration. We assume irreducibility of the matrix (m_{ij}). When $b_{ij} = 0$ for all i, j the multitype branching law of the system is critical, and when $\beta_{ij} = 1$ for all i, j it has finite second moments.

Let $N \equiv \{N(t) \equiv (N_1(t), \ldots, N_k(t)), t \geq 0\}$, where, for each i, $N_i(t) = \sum_j \delta_{x_{ij}(t)}$, and $\{x_{ij}(t)\}_j$ are the locations of the particles of type i present at time t. The process N, called the particle process, takes values in $(\mathcal{N}_p(\mathbb{R}^d))^k$, is Markovian and has a version with càdlàg paths. The Laplace functional of $N(t)$ is given by

$$E[\exp\{-\langle N(t), \Phi \rangle\} \mid N(0) = \mu] \;=\; \exp\{\langle \mu, \log(1 - u(\Phi, t)) \rangle\}, \tag{2.1}$$

$$\Phi = (\varphi_1, \ldots, \varphi_k) \in (K_p(\mathbb{R}^d))^k, \quad \mu \in (\mathcal{N}_p(\mathbb{R}^d))^k,$$

(with $\log(1 - u) = (\log(1 - u_1), \ldots, \log(1 - u_k))$), where $u(\Phi, x, t) = (u_1(\Phi, x, t), \ldots, u_k(\Phi, x, t))$, $x \in \mathbb{R}^d$, $t \geq 0$, is the unique (mild) solution of the system of non–linear p.d.e.'s

$$\frac{\partial}{\partial t} u_i \;=\; \Delta_{\alpha_i} u_i + V_i \sum_{j=1}^{k} [m_{ij}(1 + b_{ij}) - \delta_{ij}] u_j - V_i \sum_{j=1}^{k} m_{ij} c_{ij} u_j^{1+\beta_{ij}}, \tag{2.2}$$

$$u_i(x, 0) \;=\; 1 - \exp\{-\varphi_i(x)\}, \qquad i = 1, \ldots, k.$$

This is obtained by the usual renewal argument.

3 The multitype measure branching process

The multitype measure branching process X described below is a rescaling limit of the particle process N defined in Section 2, such that the particles have a small mass and a short lifetime, and the mean mass of the particles in any bounded set remains bounded.

For each $n = 1, 2, \ldots$, let N^n denote the particle process with parameters V_i^n, m_{ij}^n, b_{ij}^n and c_{ij}^n such that $V_i^n = V n^{\beta_{ii}}$, and $m_{ij}^n \to m_{ij}$, $b_{ij}^n \to b_{ij}$, $c_{ij}^n \to c_{ij} > 0$ as $n \to \infty$ in such a way that

$$n^{\beta_{ii}}[m_{ij}^n(1 + b_{ij}^n) - \delta_{ij}] \to \theta_{ij}$$

and

$$m_{ij}^n c_{ij}^n n^{\beta_{ii} - \beta_{ij}} \to \gamma_{ij}.$$

It follows that $m_{ij} = \delta_{ij}$, $b_{ii} = 0$, $\gamma_{ij} = 0$ for $i \neq j$, $\gamma_i \equiv \gamma_{ii} = c_{ii}$, $\theta_{ij} \geq 0$ for $i \neq j$, and θ_{ii} can be negative ($\sum_{j=1}^k \theta_{ij} = 0$ in the critical case). In particular, in this rescaling each particle produces, in the limit, only particles of its own type and in a critical way, and the only remnant of the change of types is contained in the constants θ_{ij}, $i \neq j$. In addition, in the n–th rescaling each particle has a mass n^{-1}. Let $X^n = n^{-1} N^n$ denote the mass process.

The following theorem can be proved using the rescaled versions of (2.1) and (2.2), similarly to the well–known results on the subject (see e.g. Dawson, 1993, Gorostiza and López–Mimbela, 1990, 1993).

Theorem 3.1 *If $X^n(0) \Rightarrow X(0)$ as $n \to \infty$, then $X^n \Rightarrow X$ in $D([0, \infty), (\mathcal{M}_p(\dot{\mathbb{R}}^d))^k)$ as $n \to \infty$, where X is a Markov process with values in $(\mathcal{M}_p(\mathbb{R}^d))^k$ and Laplace functional given by*

$$E[\exp\{-\langle X(t), \Phi \rangle\} \mid X(0) = \mu] \;=\; \exp\{-\langle \mu, u(\Phi, t) \rangle\}, \tag{3.1}$$

$$\Phi = (\varphi_1, \ldots, \varphi_k) \in (K_p(\mathbb{R}^d))^k, \mu \in (\mathcal{M}_p(\mathbb{R}^d))^k,$$

and $u(\Phi, x, t) = (u_1(\Phi, x, t), \ldots, u_k(\Phi, x, t))$, $x \in \mathbb{R}^d$, $t \geq 0$, is the unique (mild) solution of the system of non–linear p.d.e.'s

$$\frac{\partial}{\partial t} u_i \;=\; \Delta_{\alpha_i} u_i + V_i \sum_{j=1}^k \theta_{ij} u_j - V_i \gamma_i u_i^{1+\beta_{ii}}, \tag{3.2}$$

$$u_i(x, 0) \;=\; \varphi_i(x), \qquad\qquad i = 1, \ldots, k.$$

In the system (3.2) the i–th equation contains only the non–linear term in u_i, in contrast to the system (2.2) related to the particle process, where all the equations contain non–linear terms in all the components. This is due to the fact that in the rescaling limit each particle produces only particles of its own type, and it is consistent with the structure of the branching mechanism for multitype continuous–state branching processes given in Rhyzhov and Skorokhod (1970), and Watanabe (1969).

The case $\beta_{ij} = 1$ for all i, j is covered in detail in Gorostiza and López–Mimbela (1990), where a martingale characterization of X is also given. In this case, and only in this case, the process X has continuous paths, and some of its properties are described in Gorostiza and Roelly (1991), including results on the occupation time process, a stochastic evolution equation for X with a martingale measure driving process, the existence of a random density, and the Hausdorff dimension of a random support. Concerning the latter, for each $i = 1, \ldots, k$, the support of the i–th component of X has Hausdorff dimension equal to $\min\{\alpha_i, d\}$ a.s., which is the same as if there was no interaction of types (i.e., $\theta_{ij} = 0$ for $i \neq j$). This is also a consequence of the fact that in the rescaling limit each particle produces only particles of its own type. An interesting question is to describe the interaction between the random supports of the different components.

4 Long time behavior of critical branching systems and related non–linear partial differential equations

As noted above, the systems of p.d.e.'s related to the particle systems allow a wider class of non–linear terms than those related to the superprocess limit. In the critical case the long time behavior of the solution of the system (2.2) can be investigated by means of the particle process.

Let $N \equiv \{N(t) \equiv (N_1(t), \ldots, N_k(t)), \ t \geq 0\}$ be the particle process defined in Section 2, in the critical case (i.e., $b_{ij} = 0$ for all i, j). We will describe some results on the equilibrium states of N, criteria for persistence (i.e., non–triviality of the equilibrium state), extinction, and convergence to equilibrium as $t \to \infty$, and corresponding results on the long time behavior of the system of non–linear p.d.e.'s (2.2). The references for these results are Gorostiza, Roelly and Wakolbinger (1992), and Gorostiza and Wakolbinger (1991,1992,1993,1994).

Let $(\mathcal{U}_t)_{t \geq 0}$ denote the semigroup (on $(C(\mathbb{R}^d))^k$) generated by the unbounded operator $\mathcal{A}$ defined by

$$(\mathcal{A}\Phi)_i(x) = \Delta_{\alpha_i} \varphi_i(x) + V_i \sum_{j=1}^{k} (m_{ij} - \delta_{ij}) \varphi_j(x), \quad i = 1, \ldots, k,$$

$$\Phi = (\varphi_1, \ldots, \varphi_k).$$

It represents of the "basic process" of the system, i.e., the migration of the particles according to symmetric α_i–stable processes and the change of types with probabilities

m_{ij}, without the branching. Let $\Gamma = (\gamma_1, \ldots, \gamma_k)$, $\gamma_1, \ldots, \gamma_k > 0$, $\sum_{i=1}^{k} \gamma_i = 1$, be the normalized solution of the system

$$\sum_{j \neq i} \gamma_j V_j m_{ji} = \gamma_i V_i (1 - m_{ii}), \quad i = 1, \ldots, k.$$

Γ is an invariant measure for the Markov chain of types component of the basic process. The measure Λ_I on $\mathbb{R}^d \times \{1, \ldots, k\}$ defined by

$$\Lambda_I = \lambda \otimes \Gamma = (\lambda \gamma_1, \ldots, \lambda \gamma_k),$$

where λ is the Lebesgue measure on $\mathbb{R}^d$, is invariant for the semigroup $\mathcal{U}_t$. It is shown that if $N(0)$ is a Poisson random field with intensity measure Λ_I, then $N(t)$ has the same intensity $EN(t) = \Lambda_I$ for all $t \in [0, \infty)$, and $N(t) \Rightarrow N(\infty)$ as $t \to \infty$, where $N(\infty)$ is an equilibrium state for N. It follows that $EN(\infty) \leq \Lambda_I$ (component-wise). If $EN(\infty) = \Lambda_I$, the process N is said to be persistent for Λ_I. It is shown that either persistence holds or $N(t)$ goes to (local) extinction (i.e., the number of particles in any bounded set converges to 0 in probability as $t \to \infty$). In the case of persistence, $N(\infty)$ is called a Poisson–type equilibrium measure with intensity Λ_I.

The following theorem gives criteria for persistence, convergence to equilibrium, and extinction. The results depend on the critical dimension d_c defined by

$$d_c = \min \alpha_i / \min \beta_{ij}.$$

Theorem 4.1

a. If $d > d_c$, then N is persistent for Λ_I.

b. If $d > d_c$ and the initial particle configuration $N(0)$ is such that

$$\langle N(0), \mathcal{U}_t \Phi \rangle \to \langle \Lambda_I, \Phi \rangle \text{ in probability as } t \to \infty$$

for all $\Phi \in (Q_p(\mathbb{R}^d)_+)^k$, then $N(t) \Rightarrow N(\infty)$ as $t \to \infty$, where $N(\infty)$ is the Poisson–type equilibrium measure with intensity Λ_I.

c. If $d \leq d_c$, then N goes to extinction.

Parts a and b follow from general criteria for persistence and convergence to equilibrium of critical branching particle systems on Banach spaces, and part c is proved by a "backward tree" method.

The persistence/extinction result can be heuristically interpreted as follows. Due to the irreducibility of the matrix (m_{ij}), all the types are interconnected in a recurrent way. Hence the overall mobility of the system is dominated by the most mobile type, and the overall tendency to extinction is dominated by the type which goes to extinction the fastest. The most mobile type corresponds to $\min \alpha_i$ and the fastest tendency to extinction corresponds to $\min \beta_{ij}$. Due to criticality of the branching, any bounded subset of $\mathbb{R}^d$ will become empty of particles in finite time, unless the mobility of the system is large enough so that particles coming from other regions can reach the subset before dying. If $\min \alpha_i$ is small enough relative to d, the tendency to extinction is counteracted by the mobility, and persistence holds.

A large class of inittal configurations $N(0)$ which satisfy the condition in b is identified in Stöckl and Wakolbinger (1991) (see also Theorem 4 of Gorostiza and Wakolbinger, 1992).

We see from (2.1) that the long time behaviors of the process N and the solution u of the system of non–linear p.d.e.'s (2.2) should be related (in the critical case). This is in fact so, and moreover the result can be extended to the following system of non–linear p.d.e.'s.

Let $u(\Phi) = (u_1(\Phi, x, t), \ldots, u_k(\Phi, x, t))$, $x \in \mathbb{R}^d$, $t \geq 0$, be the (mild) solution of the system

$$\frac{\partial}{\partial t} u_i = \Delta_{\alpha_i} u_i + V_i \sum_{j=1}^{k} (m_{ij} - \delta_{ij}) u_j - V_i \sum_{j=1}^{k} m_{ij} c_{ij} u_j^{1+\beta_{ij}}, \qquad (4.1)$$

$$u_i(x, 0) = \varphi_i(x), \qquad\qquad i = 1, \ldots, k.$$

where $\Phi = (\varphi_1, \ldots, \varphi_k) \in (C_c(\mathbb{R}^d)_+)^k)$, with $\alpha_i \in (0, 2]$, $V_i > 0$, $\beta_{ij} \in (0, 1]$, $c_{ij} \in [0, 1/(1 + \beta_{ij})]$, and (m_{ij}) is an irreducible non–negative matrix with maximal eigenvalue 1. This system has a unique global solution, and it remains non–negative for all $t \geq 0$ (see e.g. Pazy, 1983).

Theorem 4.2 *If at least one φ_i is not 0, then for each $i = 1, \ldots, k$,*

$$\lim_{t \to \infty} \|u_i(\Phi, t)\|_{L^1} \begin{cases} > 0 & \text{if } d > d_c, \\ = 0 & \text{if } d \leq d_c. \end{cases}$$

(here $d_c = \min \alpha_i / \min\{\beta_{ij} \mid c_{ij} > 0\}$).

For non–linear heat equations on $\mathbb{R}^d$, which correspond to $k = 1$, $\alpha = 2$, i.e.,

$$\frac{\partial}{\partial t} u = \Delta u - \gamma u^{1+\beta}, \qquad (\gamma > 0),$$

this result has been proved using analytical methods by Escobedo and Kavian (1987), and Gmira and Veron (1984). We do not know of any analytical proof of Theorem 4.2.

The multitype measure–branching process X described in Section 3 can also be used to prove Theorem 4.2 in the case $c_{ij} = 0$ for $i \neq j$, using (3.1) and (3.2) with

$\theta_{ij} = m_{ij} - \delta_{ij}$ (see Gorostiza and Wakolbinger, 1992). This restriction is due to the fact explained in Section 3 about the rescaling.

5 The demographic variation process

The demographic variation process was introduced in Gorostiza (1988) as a way of measuring the changes in a branching particle system with respect to the population where there are no branchings and deaths. It is a non–Markovian signed measure valued process. Laws of large numbers and fluctuation limits under several rescalings for the demographic variation process were proved in Gorostiza (1988), and the results were extended to the multitype case in Gorostiza and López–Mimbela (1992). The main question is when the Markov property holds for the limit fluctuation process.

In the monotype case, a superprocess–type limit for the demographic variation process was obtained in Gorostiza and Roelly (1992). It is a Markovian signed measure–valued process X whose characteristic functional is given by

$$E[\exp\{-\langle X(t), \varphi \rangle\} \mid X(0) = \mu] = \exp\{i\langle \mu, u(\varphi, t) \rangle\},$$

where $u(\varphi, x, t)$ is the solution of a non–linear p.d.e. of the form

$$\frac{\partial}{\partial t} u = \Delta_\alpha u + V b(u + T\varphi) - (-i)^\beta V c(u + T\varphi)^{1+\beta},$$
$$u(0) = 0,$$

$(T_t)_{t \geq 0}$ is the semigroup generated by Δ_α, and $V > 0$, $b \in \mathbb{R}$, $c > 0$. This result can also be extended to the multitype case.

Appendix

Notation

$C(\mathbb{R}^d)$: bounded continuous functions on $\mathbb{R}^d$,

$C_c(\mathbb{R}^d)$: subset of $C(\mathbb{R}^d)$ of functions with compact support,

$\varphi_p(x) = (1 + \|x\|^2)^{-p}$, $x \in \mathbb{R}^d$, $(p > 0)$,

$K_p(\mathbb{R}^d) = C_c(\mathbb{R}^d)_+ \cup \{\varphi_p\}$,

$C_p(\mathbb{R}^d) = \{\varphi \in C(\mathbb{R}^d) : \sup_x |\varphi(x)/\varphi_p(x)| < \infty\}$,

$Q_p(\mathbb{R}^d) = \{\varphi \in C_p(\mathbb{R}^d) : \lim_{|x| \to \infty} \varphi(x)/\varphi_p(x) \text{ exists }\}$,

$\langle \mu, \varphi \rangle = \int_{\mathbb{R}^d} \varphi \, d\mu$,

$\mathcal{M}_p(\mathbb{R}^d)$: non–negative Radon measures μ on $\mathbb{R}^d$ such that $\langle \mu, \varphi_p \rangle < \infty$,

$\mathcal{N}_p(\mathbb{R}^d)$: subset of $\mathcal{M}_p(\mathbb{R}^d)$ of point measures,

$\mathcal{M}_p(\mathbb{R}^d)$ is equipped with the topology defined by the elements of $K_p(\mathbb{R}^d)$,

$\mathcal{M}_p(\dot{\mathbb{R}}^d)$: extension of $\mathcal{M}_p(\mathbb{R}^d)$, where $\dot{\mathbb{R}}^d = \mathbb{R}^d \cup \{\infty\}$ (∞ isolated point),

$D([0,\infty), \mathcal{M}_p(\dot{\mathbb{R}}^d))$: càdlàg functions from $[0,\infty)$ into $\mathcal{M}_p(\dot{\mathbb{R}}^d)$ with a Skorokhod–type topology,

$\Delta_\alpha \equiv -(-\Delta)^{\alpha/2}$, $\alpha \in (0,2]$, Δ=Laplacian on $\mathbb{R}^d$, (Δ_α is the infinitesimal generator of the symmetric stable process with index α),

p is chosen so that $p > d/2$, and in addition $p < (d+\alpha)/2$ in case $\alpha < 2$.

Obvious extensions are defined for Cartesian products (k times).

$\Rightarrow$: weak convergence of random elements.

Acknowledgement. I thank CIMAT for hospitality. This work was partially supported by CONACyT grant 2059–E9302.

References

Asmussen, S. and Hering, H. (1983), *Branching Processes*, Birkhäuser, Boston.

Athreya, K. B. and Ney, P. E. (1972), *Branching Processes*, Springer–Verlag, Berlin.

Dawson, D. A. (1992), Infinitely divisible random measures and superprocesses, in Stochastic Analysis and Related Topics, Silivri, 1992, Progr. Prob. 32, 1–129, Birkhäuser, Boston.

Dawson, D. A. (1993), Measure–Valued Markov Processes, École d'Été de Probabilités de Saint Flour 1991. *Lect. Notes Math.* **1541**, 1–260, Springer–Verlag, Berlin.

Dynkin, E. B. (1991), Branching particle systems and superprocesses, *Ann. Prob.* **19**, 1157–1194.

Escobedo, M. and Kavian, O. (1987), Asymptotic behavior of positive solutions of a nonlinear heat equation, *Houston J. Math.* **13**, 39–50.

Gmira, A. and Veron, L. (1984), Large time behavior of the solution of a semilinear parabolic equation in $\mathbb{R}^n$, *J. Diff. Equations* **53**, 259–276.

Gorostiza, L. G. (1988), The demographic variation process of branching random fields, *J. Multiv. Anal.* **25**, 174–200.

Gorostiza, L. G. and López–Mimbela, J. A. (1990), The multitype measure branching process, *Adv. App. Prob.* **22**, 49–67.

Gorostiza, L. G. and López–Mimbela, J. A. (1992), The demographic variation process of multitype branching random fields, *J. Multiv. Anal.* **41**, 102–116.

Gorostiza, L. G. and López–Mimbela, J. A. (1993), A convergence criterion for measure–valued processes, and application to continuous superprocesses, in *Barcelona Seminar on Stochastic Analysis*, St. Feliu de Guíxols, 1991, (Nualart, D. and Sanz Solé, M., Editors) Progr. Prob. 32, 62–71, Birkhäuser, Boston.

Gorostiza, L. G. and Roelly, S. (1991), Some properties of the multitype measure branching process, *Stoch. Proc. Appl.* **37**, 259–274.

Gorostiza, L. G. and Roelly, S. (1992), Superprocesos signados y una descomposición de los superprocesos de Dawson–Watanabe, *Aport. Mat., Notas Invest.* 7, 77–88.

Gorostiza, L. G., Roelly, S. and Wakolbinger, A. (1992), Persistence of critical multitype particle and measure branching processes, *Prob. Theory Relat. Fields 92*, 313–335.

Gorostiza, L. G. and Wakolbinger, A. (1991), Persistence criteria for a class of critical branching particle systems in continuous time, *Ann. Probab. 19*, 266–288.

Gorostiza, L. G. and Wakolbinger, A. (1992), Convergence to equilibrium of critical branching particle systems and superprocesses, and related nonlinear partial differential equations, *Acta Appl. Math. 27*, 269–291.

Gorostiza, L. G. and Wakolbinger, A. (1993), Asymptotic behavior of a reaction–difussion system. A probabilistic approach. *Random and Comp. Dynamics 1*, 445–463.

Gorostiza, L. G. and Wakolbinger, A. (1994), Long time behavior of critical branching particle systems and applications, in *Conference on measure–valued processes, stochastic partial differential equations and interacting systems, CRM Proc. & Lect. Notes 5*, (Dawson, D.A., Editor), Amer. Math. Soc.

Jagers, P. (1975), *Branching Processes with Biological Applications*, Wiley, New York.

Li, Z.-H. (1992), A note on the multitype measure branching process, *Adv. Appl. Prob. 24*, 496–498.

López–Mimbela, J. A. (1992), Fluctuation limits of multitype branching random fileds, *J. Multiv. Anal. 40*, 56–83.

Mode, C. J. (1971), *Multitype Branching Processes*, American Elsevier, New York.

Pazy, A. (1983), *Semigroups of Linear Operators with Applications to Partial Differential Equations*, Springer–Verlag, New York.

Rhyzhov, Yu. M. and Skorokhod, A. V. (1970), Homogeneous branching processes with a finite number o types and continuously varying mass, *Teor. Verojanost. i Primenen. 15*, 704–707.

Sevastyanov, B. A. (1971), *Branching Processes* (Russian), Nauka, Moscow.

Stöckl, A. and Wakolbinger, A. (1991), Convergence to equilibrium in independent particle systems with migration and mutation, Tech. Rep. 446, Inst. Math., Univ. Linz.

Wakolbinger, A. (1994), Limits of spatial branching populations, Bernoulli (to appear).

Watanabe, S. (1969), On two dimensional Markov processes with branching property, *Trans. Amer. Math. Soc. 136*, 447–466.

REPRESENTATION OF MARKOV CHAINS AS STOCHASTIC DIFFERENTIAL EQUATIONS

Hamza K. and Klebaner F.C*

University of Melbourne

Abstract.

A representation of continuous time Markov chain as a stochastic differential equation driven by a martingale is given. To obtain it, we give simple sufficient conditions for regularity and integrability of Markov chains in terms of their infinitesimal parameters. Stochastic differential equation representation is used for obtaining growth rates. Results on asymptotic behaviour of population dependent branching processes are obtained as an application.

Key words: regularity, integrability, branching processes, population dependence.

1. Introduction.

In studies of population dependent branching processes it has become clear that it is useful to study them in the context of wider class of models described by stochastic differential or difference equations. Such an approach for processes in discrete time was used by Kuster (1985), Keller et.al. (1987), Klebaner (1989) and Kersting (1991). Keller et al. (1985) obtained a detailed analysis of possible rates of growth as well as corresponding limit laws for diffusions. Using the results for diffusions they established similar results for discrete time models. Thus, the approach of stochastic differential equations (d.e.'s) seems to be fruitful. Unfortunately such a representation is not available in general for Markov chains in continuous time due to difficulties with controlling the jumps. The paper provides conditions for representation of Markov chains as stochastic d.e.'s and an application of this approach for studying growth rates of the solutions. These results are used to obtain growth rates in population dependent branching processes.

2. Conditions for regularity and integrability of Markov chains.

Consider a continuous time, homogeneous Markov chain $\{Z_t\}$ with countable state-space which we shall take as $\mathbf{Z}$, the space of integers. For population models such as generalized branching processes the state-space is the set of non-negative integers, however the results presented here equally hold for processes on $\mathbf{Z}$. We seek a representation of the process as a sum of a zero-mean martingale that represents the noise and the deterministic part that represents the drift. Such a representation is intimately related to the regularity (finiteness on finite intervals) and integrability of the process. The process is defined by its times of jumps, $\{\tau_n\}$ ($\tau_0 = 0$, $\tau_{n+1} = \inf\{t > \tau_n; \Delta Z_t := Z_t - Z_{t-} \neq 0\}$ and $\tau_n = \infty$ if there are less then n jumps), and the sizes of its jumps,

---

*Department of Statistics, Richard Berry Building, University of Melbourne, Parkville, Victoria, AUSTRALIA 3052. Research supported by the Australian Research Council.

$\{X_n\}$ ($X_0 = 0$). By the Markov property, conditionally on $\{Z(\tau_n) = z\}$, $\tau_{n+1} - \tau_n$ is exponentially distributed with mean, say $\lambda(z)^{-1}$, and $Z(\tau_{n+1}) - Z(\tau_n)$ has a law that depends on z only, say $\pi(z, \cdot)$. We assume that there are no absorbing states so that $\lambda(z) > 0$ for all z. Theorem 1 and its corollary give two readily testable sufficient conditions, namely (A) and (B), for regularity in terms of the infinitesimal parameters, $\lambda(z)$ and the moments of $\pi(z, \cdot)$.

The well-known necessary and sufficient condition for regularity is $\sum 1/\lambda(Z(\tau_n)) = +\infty$ a.s., see e.g. Chung (1967), p.259-260. However, this condition is often hard to check as it involves the embedded chain.

Let $m(z) = \int x\pi(z, dx)$ and $|m|(z) = \int |x|\pi(z, dx)$.

Theorem 1. Assume that the process is non-negative. If

$$\lambda(z)m(z) = O(z), \quad z \to +\infty, \tag{A}$$

then

$$\tau_n \to +\infty \ a.s., \quad \text{and for all } t \ \Pr[Z_t < +\infty] = 1.$$

Proof. Without loss of generality we may assume that Z_0 is constant, otherwise condition on $Z_0 = z$. Let $\{\mathcal{F}_t\}$ be the natural filtration of the process Z_t. We have $Z(\tau_{n+1}) = Z(\tau_n) + X_{n+1}$. Let $T = \inf\{n; \ Z(\tau_n) = 0\}$. It follows from that the process, indexed by n, $Z(\tau_n) \prod_{i=0}^{n-1}[1 + m(Z(\tau_i))/Z(\tau_i)]^{-1}I(n < T)$ is a non negative $\mathcal{F}_{\tau_n}$-martingale. By the martingale convergence theorem it must converge almost surely to a finite limit. The result now follows by (A).

For general processes on $\mathbf{Z}$, by using the variation process $|Z|_t = |Z_0| + \sum_{s \le t}|\Delta Z_s|$, we have immediately the following result.

Corollary 1. Assume that

$$\lambda(z)|m|(z) = O(|z|), \quad |z| \to +\infty. \tag{B}$$

Then

$$\tau_n \to +\infty \ a.s., \quad \text{and for all } t \ \Pr[|Z_t| < +\infty] = 1.$$

We now deal with the integrability of the process Z_t, and more generally $f(Z_t)$. If f is a function vanishing at infinity such that

$$Lf(z) := \lambda(z) \int [f(y + z) - f(z)]\pi(z, dy)$$

also vanishes at infinity, then is known that the process

$$N_t^f := f(Z_t) - f(Z_0) - \int_0^t Lf(Z_s)ds$$

is a local martingale, L being the infinitesimal generator of the Markov process Z, see e.g. Ethier and Kurtz (1986), p. 376. In Theorem 2 and its corollaries, we extend the domain of the generator L, i.e. make of N_t^f a martingale, for a much wider set of functions, as well as give a sufficient condition, condition (C), for the integrability of the process $f(Z_t)$.

Theorem 2. Assume that the process Z is regular. Let f be an unbounded function such that

$$|L|f(z) := \lambda(z) \int |f(z+y) - f(z)| \pi(z, dy) = O(f(z)), \quad |z| \to +\infty. \tag{C}$$

If $f(Z_0)$ is integrable, then so is $f(Z_t)$. Moreover N^f is a martingale and the Dynkin's formula holds,

$$E[f(Z_t)] = E[f(Z_0)] - \int_0^t E[Lf(Z_s)]ds.$$

Proof. We shall only indicate the steps of the proof as it is rather technical. The reader is referred to Hamza and Klebaner (1993) for more details. In step one we establish that N^f is a local martingale. In step two we show that $f(Z_t)$ is integrable. Finally in step three we establish that N^f is a martingale.

Let $m_k(z) = \int x^k \pi(z, dx)$ and $|m|_k(z) = \int |x|^k \pi(z, dx)$. Applying the above theorem to the particular case of polynomials we get the following corollary.

Corollary 2. Assume that Z_0^k is integrable and that for all $i = 1, \ldots, k$

$$\lambda(z)|m|_i(z) = O(|z|^i), \quad |z| \to +\infty.$$

Then Z_t^k is integrable and

$$Z_t^k - Z_0^k - \sum_{i=0}^{k-1} \binom{k}{i} \int_0^t \lambda(Z_s) Z_s^i m_{k-i}(Z_s) ds$$

is a martingale.

Using Theorem 2 with the identity function we obtain the desired representation for Z_t. It is also desirable to obtain the variance of the martingale in the representation. When a second-order condition is added we have the following result. Define the operator

$$\Gamma f(z) = \lambda(z) \int (f(z+y) - f(z))^2 \pi(z, dy).$$

Theorem 3. Assume that there exists f such that
(1) $|L|f(z) = O(|f(z)|), \, z \to \infty$,
(2) $\Gamma f(z) = O(f^2(z)), \, z \to \infty$,
(3) $f^2(Z_0)$ is integrable.
Then $\langle N^f, N^f \rangle_t = \int_0^t \Gamma f(Z_s)ds$; moreover $E\langle N^f, N^f \rangle_t < \infty$.

Proof. Direct calculations show that

$$Lf^2(z) = \Gamma f(z) + 2f(z)Lf(z),$$

and by assumptions (1) and (2) of the theorem

$$|L|f^2(z) \le \Gamma f(z) + 2|f(z)||L|f(z) = O(f^2(z)).$$

Thus by Theorem 2

$$f^2(Z_t) - f^2(Z_0) - \int_0^t \Gamma f(Z_s)ds - 2\int_0^t f(Z_s)Lf(Z_s)ds$$

is a martingale. Also with $Y_t = f(Z_t)$ we have

$$[N^f, N^f]_t = \sum_{s \leq t}(\Delta N_s^f)^2 = \sum_{s \leq t}(\Delta Y_s)^2 = Y_t^2 - Y_0^2 - 2\int_0^t Y_{s-}dY_s$$

$$= f^2(Z_t) - f^2(Z_0) - 2\int_0^t f(Z_s-)dN_s^f - 2\int_0^t f(Z_s)Lf(Z_s)ds.$$

By differencing this expression with the one above we obtain that

$$[N^f, N^f]_t - \int_0^t \Gamma f(Z_s)ds$$

is a martingale. Since $\int_0^t \Gamma f(Z_s)ds$ is continuous, the first result follows. The integrability can be seen to follow from Assumption (2) and the integrability of $f^2(Z_t)$.

3. Asymptotic behaviour of solutions of perturbed linear equations.

We give results on the asymptotic behaviour of Markov processes that serve as models for populations developing in time and are therefore nonnegative. We start this section by stating the main representation by taking $f(z) = z$ in Theorem 3. Let $\sigma^2(z) = m_2(z)$. Suppose that

$(A0)$ $\qquad \lambda(z)|m|(z) = O(z), \quad and \quad \lambda(z)\sigma^2(z) = O(z^2), \quad z \to \infty.$

Then

$$Z_t = Z_0 + \int_0^t \lambda(Z_s)m(Z_s)ds + N_t, \tag{1}$$

where N_t is a martingale with $EN_t^2 = \int_0^t E(\lambda(Z_s)\sigma^2(Z_s))ds < \infty$.

The infinitesimal mean change in the process when the process is in the state z is $g(z) = \lambda(z)m(z)$, as can be seen from (1). Our main assumption (A1) is that it is asymptotically linear, namely

$(A1)$ $\qquad \lambda(z)m(z) = rz + D(z), \quad D(z) = o(z), \quad z \to \infty.$

In the context of branching models this assumption represents the stabilization of reproduction assumption, see Application to branching processes in section 4 below. Processes with asymptotically linear rate of change can be considered as perturbed linear differential equation, where there are perturbations to the dynamical deterministic part as well as random noise. Provided that both types of pertubations are not too large the solutions behave like the deterministic solutions of linear equations.

Theorem 4. (Result on exponential growth).
Suppose (A0)-(A4) hold, where

(A2) $r > 0$,

(A3) $|D(z)| \leq \delta(z)$ for a function $\delta(x)$ such that $\delta(x)/x$ is non increasing with $\int_1^\infty \delta(x)/x^2 dx < \infty$,

(A4) $\lambda(z)\sigma^2(z) \leq v(z)$ for a function $v(x)$ such that $v(x)/x^2$ is non increasing with $\int_1^\infty v(x)/x^3 dx < \infty$.

Then $Z_t e^{-rt}$ converges in L^2 and almost surely as $t \to \infty$.

Theorem 5. (Result on slow growth).

Suppose (A0), (A1) and (A5)-(A8) hold, where

(A5) $r = 0$,

(A6) $D(z) = c + o(1)$, $z \to \infty$,

(A7) $\lambda(z)\sigma^2(z) = \sigma^2 z + o(z)$, $z \to \infty$,

All absolute moments of jump distributions exist and

(A8) $\lambda(z)|m_k|(z) = O(z^k)$, and for $k = 3, 4, ... \, \lambda(z)m_k(z) = o(z^{k-1})$, $z \to \infty$.

Suppose also that $\{Z_t\}$ is transient. Then Z_t/t converges in distribution to the gamma distribution with parameters $2c/\sigma^2$ and $2/\sigma^2$ as $t \to \infty$.

Other cases of slow growth can be recovered from the following result. Let

$$G(x) = \int_1^x \frac{ds}{g(s)}.$$

Theorem 6. Suppose that (A0) and (A9)-(A11) hold, where

(A9) $G(\infty) = \infty$, $g'(x)G(x) \to c$, $x \to \infty$,

(A10) $|L|G^k(z) = O(G^k(z))$ for all $k = 1, 2, ..$,

(A11) $\lambda(z)\sigma^2(z) \sim \beta g^2(z)G(z)$, $z \to \infty$.

Then if $\beta > 0$, $G(Z_t)/t$ converges to a gamma limit with parameters $2/\beta - c$, $2/\beta$; whereas if $\beta = 0$, $G(Z_t)/t$ converges in probability to a constant.

4. Application to population dependent branching processes.

A population dependent branching process is a model where a particle in a population of size z lives an exponentially distributed lifetime with parameter $a(z)$, and at the end of its life produces a random number of offspring according to a distribution $\pi(z)$. The reproductive behaviourof different particles is independent of each other and the past. As the individuals reproduce at the end of their life, it is clear that the lifespan parameters and not only the offspring distributions determine the reproduction rate. In classical Markov branching processes $a(z) = a$ and $m(z) = m$ are constants. $r = am$ is known as the Malthusian parameter, it determines largely the asymptotic behaviour of the process. A process with $r > 0$ is known as the supercritical process; the population grows at the rate e^{rt}. If $r = 0$ then the process is known as critical, it ultimately dies out, but can grow large before that, (its mean given that the process has not died out and its second moments grow like t see e.g. Athreya and Ney (1972)). Populations with stabilizing reproduction are such that

$$a(z)m(z) \to r, \quad z \to \infty.$$

Recall that $a(z)m(z)$ is the mean change rate, see equation (1) above. In a process with stabilizing reproduction, reproduction stabilizes in the sense of the mean change rate in the population. Analogously to the classical case we call the process near-supercritical if $r > 0$ and near-critical if $r = 0$. It is easily seen that in a population dependent process the holding time at z is $\lambda(z) = za(z)$. The assumption of stabilization of reproduction yields $\lambda(z)m(z) = za(z)m(z) \sim rz$, $z \to \infty$, so that our main assumption (A1) is satisfied. In a similar way, conditions of Theorems 4-6 translate into conditions on parameters of the individuals' reproduction $a(z)$ and $m(z)$. We note that in branching models state 0 is absorbing, and limit results hold on the set $\{Z_t \to \infty\}$.

5. Remarks.

1. An example of a process satisfying the assumptions in Theorem 6 is provided by a process with $g(x)$ regularly varying at infinity with parameter $\alpha < 1$, and $|m|_k(z) = O(z^k)$ for $k > 2$, $z \to \infty$. More general functions g may be considered and moment conditions relaxed similar to Kersting (1991).

2. Kuster (1983) gave results for processes that grow fast, approximately exponentially. For processes with slow growth our results generalize those of Reinhard (1990).

3. Results similar to Theorems 4-6 hold for multitype processes; Klebaner (1993).

6. Proofs of results in section 3.

For the proof of Theorem 4 the following lemmatta are required.

Lemma 1. Let $f(x) > 0$ be non-increasing, continuous, such that $xf(x)$ non-decreasing and $\int_0^\infty f(x)/x\,dx < \infty$. Suppose $a(x) \geq 0$ is differentiable, and satisfies for $r > 0$, $|a'(x)| \leq a(x)f(a(x)e^{rx})$. Then $\lim_{x\to\infty} a(x) = a$ exists and $a > 0$ if $a(0)$ is large enough.

Lemma 2. Let $f(x) > 0$ be such that $f(x)/x$ is non-increasing to zero and $\int_1^\infty f(x)/x^2\,dx < \infty$. Then there exists $F(x)$ such that $F(x) \geq f(x)$, $F(x)/x$ is non-increasing to zero, $F(x)$ is non-decreasing and concave on R^+, and $\int_1^\infty F(x)/x^2\,dx < \infty$.

Proof of Lemma 1 is given in Klebaner (1993), and Lemma 2 is in Klebaner (1989).

Proof of Theorem 4. Using integration by parts in (1) we obtain for $W_t = Z_t e^{-rt}$,

$$W_t = W_0 + \int_0^t e^{-rs} D(Z_s)ds + \int_0^t e^{-rs} d(N_s).$$

By using the Cauchy-Schwarz inequality, (A3) and the appropriate concave function guaranteed by Lemma 2 we obtain

$$E|D(Z_t)| \leq C\delta(EZ_t)$$

where $\delta(x)/x = \delta_1(x)$ satisfies the conditions of Lemma 2 and $\int_0^\infty \delta_1(x)/x\,dx < \infty$. Let $w(t) = EW_t$. It can be seen that it satisfies conditions of Lemma 1, which gives that $\lim_{t\to\infty} w(t)$ exists. Moreover it is positive if Z_0 is taken to be a large enough constant. Let $T_t = \int_0^t e^{-rs} D(Z_s)ds$. In view of finiteness of the $\lim_{t\to\infty} w(t)$,

$$E\int_0^\infty |e^{-rs} D(Z_s)|ds < C\int_0^\infty w(s)\delta_1(e^{rs}w(s))ds < C\int_0^\infty \delta_1(e^{rs})ds < \infty.$$

This implies that T_t converges to T_∞ a.s. and in L^1, as $t \to \infty$. As $W_t \geq 0$, $E|W_t| = EW_t$ which we showed is bounded. $E|T_t|$ is also bounded. Almost sure convergence follows by the martingale convergence theorem. Similar arguments regarding second moments show that W_t converges in L^2.

Proof of Theorem 5. The imposed conditions assure that polynomials satisfy conditions of Theorem 2 so that Z_t^k is integrable for all k. Taking expectations in (1) we obtain $EZ_t = EZ_0 + \int_0^t ED(Z_s)ds$. ¿From here and (A6) it follows that $EZ_t \sim ct$. Let $h_k(t) = E(Z_t^k)$; then it follows that

$$h_k(t) = h_k(0) + c_k \int_0^t h_{k-1}(s)ds + \int_0^t o(h_{k-1}(s))ds,$$

with $c_k = ck + \binom{k}{2}\sigma^2$. As $h_1(t) \sim ct$, it now follows by induction that $h_k(t) \sim \beta_k t^k$, where β_k satisfy the recurrence relation

$$\beta_k = (c + \sigma^2(k-1)/2)\beta_{k-1}.$$

This relation together with the initial condition determines the moments of the required distribution.

Proof of Theorem 6. Taylor's formula is applied to $G^k(Z_{t+s}) - G^k(Z_t)$ to set up equations for $h_k(t) = EG^k(Z_t)$. As it directly follows the steps of the previous proof, although technically more complicated, we do not give details here.

References

[1] Chung K.L. (1967) *Markov Chains with Stationary Transition Probabilities*, Springer, Berlin.

[2] Ethier S.N., Kurtz T.G. (1986) *Markov Processes*, Wiley, New-York.

[3] Hamza K., Klebaner F.C. (1993) Conditions for regularity and integrability of Markov Chains. Uni. Melb. Dept. Stats. Res. Rep No 8.

[4] Jacod J., and Shiryaev A.N. (1987) *Limit Theorems for Stochastic processes*, Springer, Berlin.

[5] Keller G., Kersting G. and Rosler U. (1984) On the asymptotic behaviour of solutions of stochastic differential equations. *Z. Wahrscheinlichkeitstheorie* 68, 163-189.

[6] Keller G., Kersting G. and Rosler U. (1987) On the asymptotic behaviour of discrete time stochastic growth processes. *Ann. Probab.* 15, 305-343.

[7] Kersting G. (1990) Some properties of stochastic difference equations. In: *Stochastic Modelling in Biology*. P. Tautu Editor, 328-339, World Scientific, Singapore.

[8] Kersting G. (1991) A Law of Large Numbers for stochastic difference equations. *Stoch. Proc. Appl.*, 40, 1-14.

[9] Kersting G. (1991) Asymptotic Gamma distribution for stochastic difference equations. *Stoch. Proc. Appl.*, 40, 15-28.

[10] Klebaner F.C. (1989) Stochastic difference equations and generalized gamma distributions. *Ann. Probab.* 17, 178-188.

[11] Klebaner F.C. (1989) Geometric growth in near-supercritical population size dependent multitype branching processes. *Ann. Probab.* 17, 1466-1477.

[12] Klebaner F.C. (1994) Asymptotic behaviour of Markov population processes with asymptotically linear rate of change. *J. Appl. Probab.* to appear.

[13] Kuster P. (1983) Generalized Markov branching processes with state-dependent offspring distributions. *Z. Wahrscheinlichkeitstheorie* 64, 475-503.

[14] Kuster P. (1985) Asymptotic growth of controlled Galton-Watson processes. *Ann. Probab.* 13, 1157-1178.

[15] Reinhard I. (1990) The qualitative behaviour of some slowly growing population-dependent Markov branching processes. In: *Stochastic Modelling in Biology.* P. Tautu Editor, 267-277, World Scientific, Singapore.

AN EXTENSION OF A GALTON-WATSON PROCESS TO A TWO-SEX DENSITY DEPENDENT MODEL.

Charles J. Mode
Drexel University, Philadelphia, USA

Abstract

A two-sex, density dependent, genetic model is formulated as an extension of a multitype Galton-Watson process to a non-linear case. The formulation presented and partially analyzed in this paper is new in at least three ways to the field of branching processes. First, a new parameterization of the mating system is presented based on, among other things, the Farlie-Morgenstern class of bivariate distribution functions. Second, the viabilities of genotypes are formulated as a function of total population size, which allows for accommodating density dependence into the model to take into account the carrying capacity of an environment. Third, a set on nonlinear difference equations is embedded in the stochastic process by operating with conditional expectation of the random functions of the process at each point in time, given the past, to obtain functions which are viewed as estimates of the sample functions of the process. These equations resemble a model that would arise if one worked within deterministic paradigm. Computer intensive methods are used to partially analyze the model by computing a sample of Monte Carlo realizations of the process, summarizing them statistically, and comparing the results to the predictions based on the embedded deterministic model. The results of two computer experiments are presented, which suggest that numerical solutions of the deterministic model are not always good measures of central tendency for the sample functions of the process, particularly when the deterministic model exhibits chaotic behavior or when the carrying capacity of the environment forces the numbers of some genotypes to be small. The results of these experiments suggest that density dependent formulations in which population size is limited by the carrying capacity of the environment may lead to alternative working paradigms for the study of genetic drift, a subject introduced into evolutionary genetics by two founding fathers Fisher and Wright.

Key Words: Genotype, Autosomal, Computer Intensive Methods, Monte Carlo Methods, Density Dependence, Non-linear Multitype Galton-Watson Process.

1   Introduction

From the point of view of the dynamics of biological populations, models based on branching processes have been considered incomplete, because such factors as two sexes, various types of mating systems, several genotypes in case of population genetics, and density dependence have not, for the most part, been incorporated into branching processes. The primary reason for not incorporating such factors into population models based on past formulations of branching processes was that they gave rise to non-linearities which were difficult to handle with the analytic tools, such as systems of renewal type integral equations and matrices with non-negative elements, commonly used in analyzing these classes of multitype processes. Today, however, thanks to powerful desk-top computers, non-linear models can be analyzed numerically. The purpose of this paper is to formulate and partially analyze a two-sex density dependent genetic model, which, twenty-five years ago, seemed out of reach of methods of analysis then available. In the literature on branching processes, there are relatively few papers on two-sex models. The book [1] contains an analysis of a two-sex model, which is an extension of the Galton-Watson process, and another paper on the subject is [2] .

The plan of the paper is as follows. Section 2 contains a new parametric formulation of the mating system designed to accommodate random as well as various schemes of positive and negative assortative mating, corresponding to choices of parameter values. The genetic principles underlying the model are presented in Section 3. Section 4 is devoted to a description of the stochastic population process, which forms a firm mathematical basis for Monte Carlo simulations and extends the classical two-type Galton-Watson process to accommodate two sexes, general types of mating systems, density dependence, and the genetics of a population with respect to one autosomal locus. In Section 5, a set of embedded non-linear difference equations are derived by operating on conditional expectations of the present, given the past, in a manner analogous to techniques used for prediction in time series. The paper concludes in Section 6 with the graphical presentation of the results of two computer experiments.

2 A Parameterization of the Mating System

A basic problem in formulating a two-sex branching process with several types for each sex is that of parameterizing the mating system. Mathematically, the problem may be stated as follows: Given two distribution functions F(x) and G(y), construct a bivariate distribution function H(x,y) of two random variables X and Y such that F(x) and G(y) are the marginal distribution functions, with independence of X and Y as a special case. Among the authors who have discussed this problem are Mardia [3] , Whitt [4] , and Johnson and Kotz [5] . For the class of problems under consideration, it will suffice to restrict attention to the case in which the random variables X and Y have finite variances. In this case, the distribution function that minimizes the correlation between X and Y is

$$H_o(x,y) = \max(0, F(x) + G(y)) ,\tag{1}$$

and the distribution function that maximizes the correlation between X and Y is

$$H_1(x,y) = \min(F(x), G(y)) .\tag{2}$$

These results are frequently attributed to Hoeffding and Frechet. Another class of bivariate distribution function with the marginals F(x) and G(y) is that determined by the Farlie-Morgenstern formula

$$H_2(x,y) = F(x)G(y)\left[1 + \alpha(1 - F(x))(1 - G(y))\right] , \tag{3}$$

where α is a parameter such that $|\alpha| \leq 1$. Observe that when $\alpha = 0$, the random variables X and Y are independent and the cases $\alpha = 1$ and $\alpha = -1$ correspond, respectively, to positive and negative correlations of the random variables X and Y.

A limitation of the Farlie-Morgenstern class of distributions is that the correlation ρ between the random variables X and Y is necessarily rather low. For example, for the case F and G are the uniform distribution function on the interval $[0,1]$, it can be shown that $|\rho| \leq 1/3$; moreover, numerical experiments with other marginals suggest that the Farlie-Morgenstern system by itself would be insufficient to characterize various systems of assortative mating in which one would expect values of ρ such that $|\rho| \geq 1/3$. One is thus led to consider mixtures of the three distribution functions described above to obtain higher levels of correlation. Let θ_0 and θ_1 be numbers such that $0 \leq \theta_0 \leq 1$, $0 \leq \theta_1 \leq 1$, $0 \leq \theta_0 + \theta_1 \leq 1$, and let $\theta_2 = 1 - \theta_0 - \theta_1$.Then, the mixture

$$H(x,y) = \theta_0 H_0(x,y) + \theta_1 H_1(x,y) + \theta_2 H_2(x,y) , \tag{4}$$

is a bivariate distribution function with marginals F(x) and G(y). As will be demonstrated in subsequent sections, by varying the three parameters θ_0, θ_1, and α, it will be possible to create a variety of mating systems, including random and various schemes of assortative mating.

3   Genetics and Offspring Distributions

An excellent introduction to the principles of general genetics is the book [6]; the book [7] is devoted to theories of population genetics, and [8] is an extensive treatise on the genetics of human populations. Consider a diploid species, such as man, in which all chromosomes occur in pairs so that each offspring inherits one set of chromosomes from his mother and one from his father. Among the simpler genetic situations is that of considering a population with respect to one autosomal locus with two alleles A and a . In this case, there are three genotypes for both sexes symbolized by AA, Aa, and aa, which will be numbered 1,2, and 3. A mating will be said to be of type $\langle i,j \rangle$ if the female is of genotype i and the male is of genotype j, where i, j = 1,2, and 3, giving rise to nine types of matings. In a two-sex species, all offspring in a given generation are produced from the matings in the previous generation. In a natural population, the number of individuals of each genotype in each sex may affected by differential rates of reproduction among the nine types of matings. Accordingly, let $p(i,j;n), n = 0,1,2,\cdots$,be the probability density function, (p.d.f.) of the total number of offspring produced by a mating of type $\langle i,j \rangle$ per generation, let $N(i,j)$ be a random variable with this density, and let $\mu(i,j)$ be the finite expectation of $N(i,j)$.

Offspring from any mating will be classified by sex and genotype and will be represented by the symbol $\langle k, l \rangle$, where k represents the sex of the offspring and l the genotype. When k = 1, the offspring is female and k = 2 stands for a male. For a mating of type $\langle i, j \rangle$, let q(i,j;k,l) be the probability that an offspring is of type $\langle k, l \rangle$. Because an autosomal locus is under consideration, sex and genotype are inherited independently. Let p_f be the probability an offspring is female and let $p_m = 1 - p_f$ be the probability an offspring is male. Values of the probability distribution q(i,j;k,l) vary with the type of mating $\langle i, j \rangle$.

By way of illustration, suppose the mating is of type $\langle Aa, Aa \rangle$. Then in the absence of mutation, the gametes of the female and male carry the genes A and a with probability 1/2. Therefore, among female offspring, the three genotypes AA, Aa, and aa occur, respectively, with probabilities $p_f(1/4, 1/2, 1/4)$; a similar expression may be written down for the male offspring. By continuing in this way, the six-dimensional distributions corresponding to the nine types of matings could be written down but the details will be omitted. Let W(i,j;k,l) be a random variable representing the number of offspring of type $\langle k, l \rangle$ from a mating of type $\langle i, j \rangle$ per generation, and let $\mathbf{W(i,j)}$ be a six-dimensional random vector with components W(i,j;k,l) for k = 1,2 and l = 1,2, 3. Then, given and value of the random variable $N(i, j)$, the random vector $\mathbf{W(i,j)}$ has a conditional multinomial distribution with index $N(i, j)$ and probabilities q(i,j;k,l), where k = 1 2 and l = 1, 2 , 3. In all computer simulations reported in this paper, the p.d.f. $p(i, j; n), n = 0, 1, 2, \cdots$, of $N(i, j)$ was chosen as a Poisson with parameter $\mu(i, j)$ and it was assumed that the these random variables were independent among matings types.

4   A Stochastic Population Process

Just as in Galton-Watson processes, time $t = 0, 1, 2, \cdots$ will correspond to generations and the primary focus of attention will be random functions representing the number of females and males of each genotype who enter into matings that contribute offspring to the next generation. Let $X(t, i)$ and $Y(t, i)$, respectively, be the number of such females and males of genotype $i = 1, 2, 3$ at time t. The total number of such females at time t is

$$X(t, \cdot) = \sum_{i=1}^{3} X(t, i) \tag{5}$$

and the random function $Y(t, \cdot)$ is defined similarly for males. For $X(t, \cdot) > 0$ and $Y(t, \cdot) > 0$, the relative frequencies of the genotype i among females and males are given by $u_f(t, i) = X(t, i)/X(t, \cdot)$ and $u_m(t, i) = Y(t, i)/Y(t, \cdot)$. The distribution function determined by these frequencies for females is

$$U_f(t, i) = \sum_{\nu=1}^{i} u_f(t, \nu) \ 1 \le i \le 3 \tag{6}$$

and the distribution function $U_m(t, i) \ i = 1, 2, 3$ is determined similarly for males.

At time t, there will be some total number of matings that may contribute offspring to the next generation; let $N(t)$ be this random function. Various choices for this random function have been used in literature on two-sex branching process, but in this paper this function will be chosen as

$$N(t) = [\gamma \min(X(t,\cdot), Y(t,\cdot))] , \qquad (7)$$

where $[\cdot]$ stands for the greatest integer function and γ is a parameter satisfying the condition $0 < \gamma \leq 1$. Among the total number of matings at time t, let the random function $Z(t; i, j)$ be the total number of matings of type $\langle i, j \rangle$. The conditional distribution of this random function, given $N(t)$, will be determined as follows.

Use the distribution functions $U_f(t, i)$ and $U_m(t, i)$ as marginals to determine a bivariate mixture distribution function $H(t; i, j)$ as described in equation (4) for given values of the parameters $\theta_0, \theta_1,$ and α. A bivariate density is then determined according to the difference formula

$$h(t; i, j) = H(t; i, j) - H(t; i - 1, j) - H(t; i, j - 1) + H(t; i - 1, j - 1) , \qquad (8)$$

where by definition $H(t; i, j) = 0$ if either $i < 0$ or $j < 0$. Then, given a value of the random variable $N(t)$, the conditional p.d.f. of the random variables $Z(t; i, j)$, for $i, j = 1, 2, 3$, is a multinomial with index $N(t)$ and probabilities $h(t; i, j)$ for $i, j = 1, 2, 3$.

Because the total number of matings containing a female of genotype i or a male of genotype j cannot exceed the total numbers $X(t, i)$ and $Y(t, j)$ of females and males of these genotypes at time t , it follows that the inequalities

$$\sum_{\nu=1}^{3} Z(t; i, \nu) \leq X(t, i) \qquad (9)$$

and

$$\sum_{\nu=1}^{3} Z(t; \nu, j) \leq Y(t, j) \qquad (10)$$

must be satisfied for all t, i and j with probability one. To ensure these inequalities are satisfied in Monte Carlo simulations, a rejection method is used, i.e., if for a given set of random numbers these inequalities do not hold, another set is computed and this process is continued until the desired inequalities do hold. The software has been so designed that in any simulation the number of rejections is counted. Mathematically, the acceptance distribution of the Z-random functions in (9) and (10) is the multinomial described above conditioned on these inequalities being satisfied.

As one might expect, the number of rejections is very sensitive to the choice of mating system. For example, a random mating system arises when $\theta_0 = \theta_1 = 0$ and the Farlie-Morgenstern parameter $\alpha = 0$. In order to avoid large numbers of rejections, which greatly lengthen the time to complete a computer simulation, values of the parameter γ, controlling the total number of matings, have been chosen in the range $0.65 \leq \gamma \leq 0.95$
Non-random mating systems, on the other hand, usually lead to fewer rejections. A

positive assortative mating system arises, for example, when $\theta_0 = 0, \theta_1 = 0.75$, and $\alpha = 1$. An example of a negative assortative mating system is the case $\theta_0 = 0.75, \theta_1 = 0$, and $\alpha = -1$.

The description the stochastic population process will be complete as soon as the values of the random functions $X(t+1, i)$ and $Y(t+1, i)$ are determined for generation $t + 1$. Toward this end, let $T(t; i, j; k, l)$ be a random function, representing the total number of all offspring of type $\langle k, l \rangle$ from matings of type $\langle i, j \rangle$ at time t. Let $W_\nu(i, j; k, l)$, $\nu = 1, 2, 3, \cdots$, be a sequence of independent and identically distributed (i.i.d.) random variables whose common distribution is that of the random variable $W(i, j; k, l)$ defined in section 3. Then, the random function $T(t; i, j; k, l)$ is given by the random sum

$$T(t; i, j; k, l) = \sum_{\nu=1}^{Z(t; i, j)} W_\nu(i, j; k, l) , \tag{11}$$

where the sum is 0 if $Z(t; i, j) = 0$. The total number of all offspring of type $\langle k, l \rangle$ produced by all matings at time t is given by the random function

$$V(t; k, l) = \sum_{i=1}^{3} \sum_{j=1}^{3} T(t; i, j; k, l) . \tag{12}$$

In particular, the number of females of genotype l among the offspring at time t is $V(t; 1, l)$ and the number of males with this genotype is $V(t; 2, l)$.

Density dependence, which depends on total population size at time t, enters into the formulation through viabilities which determine the probabilities that an offspring at time t survives to enter the reproductive population in generation t + 1. Total population size at time t is given by the random function

$$T(t) = X(t, \cdot) + Y(t, \cdot) . \tag{13}$$

Let $s(t; k, l)$ be the probability that an offspring of type $\langle k, l \rangle$ produced at time t survives to enter the reproductive population in generation t + 1. These probabilities will be assumed to be of the form

$$s(t; k, l) = \delta(k, l) \exp\left[-\beta(k, l) T(t)\right] , \tag{14}$$

where the $\delta(k, l)$ is a background survival probability intrinsic to type $\langle k, l \rangle$ and $\beta(k, l)$ is a non-negative threshold parameter for density dependence for this type, reflecting carrying capacity of a given environment. By way of illustration, the smaller the value of the parameter $\beta(k, l)$ for all types of offspring $\langle k, l \rangle$., the greater the carrying capacity of the environment. In particular, when all the $\beta - parameters$ are zero, then there is no density dependence.

Having defined the survival probabilities, the values of the random functions $X(t+1, j)$ and $Y(t+1, j)$ are determined as follows: Given a value of the random variable $V(t; 1, j)$, the conditional distribution of $X(t+1,j)$, the number of females of genotype j who enter the reproductive population in generation t + 1, is binomial with index $V(t;1,j)$ and probability $s(t;1,j)$ and the conditional distribution of the random function $Y(t+1,j)$ for males is determined similarly. This completes the description of the stochastic population process. Observe that this formulation accommodates two components of natural selection; namely, differential reproduction of mating types and differential survival by sex and genotype.

5 Embedded Non-Linear Difference Equations

In classical branching processes with finite first and second moment functions, it was often straight-forward to demonstrate the these moments functions satisfied a set of linear equations. Furthermore, analyses of the asymptotic behavior of the sample functions of the process often centered on the limiting behavior of solutions to linear equations. Such techniques were used extensively, for example, in the books, [9], [10], [11], and [12], as well as in many papers written on branching processes. For the process under consideration, however, the conditional expectations of the random functions $X(t+1,j)$ and $Y(t+1,j)$, for j = 1,2,3, are non-linear functions of the sample functions at t; consequently, the unconditional expectation functions of these random functions cannot satisfy a system of linear equations. However, by utilizing a concept frequently used in time series analysis, it is possible to derive a set of non-linear difference equations which closely resemble equations that would be derived if the model were formulated from a purely deterministic perspective. As is pointed out in [13] and other books on stochastic processes, the best estimate of a random function at t + 1, in the sense of minimum error mean square, is the conditional expectation of this function, given the evolution of the process up to time t.

From now on, let $\Xi(t)$ stand for a σ-algebra in the underlying probability space induced by the evolution of the random functions of the process up to time t. The basic idea underlying the derivations of the system of non-linear difference equations is to think in terms of estimating non-linear functions of the sample functions by estimates of these random functions computed in terms of conditional expectations. Estimates of random functions will be denoted by a 'hat' symbol. Thus, because, given $\Xi(t)$, the conditional distributions of the random functions $X(t+1,j)$ and $Y(t+1,j)$ are binomials with indices $V(t;1,j)$, $V(t;2,j)$ and probabilities $s(t;1,j)$, $s(t;2,j)$, it follows that the conditional expectations of these random functions take the forms

$$E\left[X(t+1,j) \mid \Xi(t)\right] = s(t;1,j)V(t;1,j) \tag{15}$$

and

$$E\left[Y(t+1,j) \mid \Xi(t)\right] = s(t;2,j)V(t;2,j) \, . \tag{16}$$

The next step in the derivation of the embedded non-linear difference equations is the estimate the non-linear functions on the right as functions of the estimates of the sample functions at t.

Let $\widehat{X}(t,l)$ and $\widehat{Y}(t,l)$ be estimates of the sample functions $X(t,l)$ and $Y(t,l)$ at t. Then, given estimates at t, the estimates of these random functions at t + 1 would be determined by

$$\widehat{X}(t+1,l) = \widehat{s}(t;1,l)\widehat{V}(t;1,l) \tag{17}$$

and

$$\widehat{Y}(t+1,l) = \widehat{s}(t;2,l)\widehat{V}(t;2,l), \tag{18}$$

where the 'hats' stand for these non-linear functions evaluated at the estimates for time t.

Observe that each i.i.d. term on the right in equation (11) has the expectation

$$E\left[W(i,j;k,l)\right] = \mu(i,j)q(i,j;k,l) . \tag{19}$$

Therefore, from equation (12) it can be seen that the conditional expectation of the random function $V(t;k,l)$ given the past is

$$E\left[V(t;k,l) \mid \Xi(t)\right] = \sum_{i=1}^{3}\sum_{j=1}^{3} Z(t;i,j)\mu(i,j)q(i,j;k,l) . \tag{20}$$

An estimate of the random function $Z(t;i,j)$ is the conditional expectation

$$E\left[Z(t;i,j) \mid \Xi(t)\right] = N(t)h(t;i,j) . \tag{21}$$

Therefore, the estimates of the V-function are determined by

$$\widehat{V}(t;k,l) = \sum_{i=1}^{3}\sum_{j=1}^{3} \widehat{N}(t)\widehat{h}(t;i,j)\mu(i,j)q(i,j;k,l). \tag{22}$$

We thus arrive at the system of non-linear difference equations

$$\widehat{X}(t+1,l) = \widehat{s}(t;1,l)\sum_{i=1}^{3}\sum_{j=1}^{3} \widehat{N}(t)\widehat{h}(t;i,j)\mu(i,j)q(i,j;1,l) \tag{23}$$

and

$$\widehat{Y}(t+1,l) = \widehat{s}(t;2,l)\sum_{i=1}^{3}\sum_{j=1}^{3} \widehat{N}(t)\widehat{h}(t;i,j)\mu(i,j)q(i,j;2,l) , \tag{24}$$

which may be used to calculate recursive estimates of the sample functions for l = 1,2,3, given the initial estimates $\widehat{X}(0,l)$ and $\widehat{Y}(0,l)$ for l = 1,2,3. The procedure used to numerically analyze the stochastic process under consideration is to compute solutions of the difference equations in (23) and (24) for a given set of parameter values. Monte Carlo realizations of the sample functions $X(t,j)$ and $Y(t,j)$, j=1,2,3, will also be computed for these parameter values. The Monte Carlo samples of the sample functions will then be analyzed statistically and compared with estimates of the sample functions computed from the non-linear difference equations. A question of interest is: in what sense are solutions of the embedded non-linear difference equations measures of central tendency for the sample functions of the process?

Before proceeding to a numerical analysis of the model, it is of interest to observe that equations (9) and (10) are satisfied in the estimates of the sample functions of the process. For, by construction,

$$\sum_{j=1}^{3} \widehat{h}(t;i,j) = \widehat{X}(t,i)/\widehat{X}(t,\cdot) \tag{25}$$

for i = 1,2,3 and the estimate of $Z(t;i,j)$ is $\widehat{Z}(t;i,j) = \left[\gamma \min(\widehat{X}(t,\cdot), \widehat{Y}(t,\cdot)) \right] \widehat{h}(t;i,j)$. Therefore, by substituting these estimates into equation (9), it follows that

$$\sum_{j=1}^{3} \widehat{Z}(t;i,j) \le \widehat{X}(t,i) \tag{26}$$

for all t≥ 1 and i = 1,2,3. By a similar argument, it can be shown that equation (10) holds in the estimates.

6   Numerical and Graphical Examples

Equations (23) and (24) constitute a non-linear transformation of the real six-dimensional space $\Re^{(6)}$ into itself indexed by a 25-dimensional parameter space described in the formulation of the model. During the last one and one-half decades such transformations have received considerable attention in the literature. A very readable introduction to the subject is the book [14]. According to existing results in the literature, such transformations can be expected to behave in at least three ways, depending on particular regions of the parameter space. One type of behavior is that iterates of the transformation will converge to a stable fixed point; a second type is that the iterates may exhibit periodic behavior; and a third type is that commonly described as chaotic, i.e., the transformation is unstable and there are no obvious periodic points among its iterates. All three types of behavior have been found in numerical experiments with the six-dimensional non-linear equation defined by equations (23) and (24). A complete analysis of the model under consideration would entail a derivation of conditions which partition the multi-dimensional parameter space into subsets such that all parameter points in a given subset would lead to one of the three types of behavior or perhaps behaviors as yet unclassified. The execution of such an analysis appears to be a very difficult problem and in this paper we shall be content with two numerical examples of biological interest.

The two numerical examples chosen for illustration differed only with respect to the 3×3 matrix $(\mu(i,j))$, representing the expected numbers of offspring by the types of matings $\langle i, j \rangle$ per generation. All other parameters of the model were held constant. The initial number of each of the three genotypes AA, Aa, and aa in both sexes was chosen as 500 so that the total size of the initial population was 3,000. The probability of a female offspring was chosen as 100/205 and that of a male as 105/205; these values are often used in the study of human populations, The theta-parameters in equation (4) were chosen as $\theta_0 = 0$, $\theta_1 = 0.75$ and $\theta_2 = 0.25$; while the Farlie-Morgenstern parameter α was chosen as $\alpha = 1$. These parameters values determine a positive assortative mating system, i.e., there is a strong tendency for females and males of the same genotype to mate. The gamma-parameter in equation (7) was chosen as $\gamma = 0.9$. With these choices of parameters characterizing the mating system, the number of times inequalities (9) and (10) did not hold were held to acceptable levels in Monte-Carlo simulations. All delta-parameters in equations (14) had the value $\delta = 1$, indicating there was no differential survival by sex and genotype. Finally, the beta-parameters in equations (14) for density dependence had the common value $\beta = 0.001$, indicating that density dependence impacted all genotypes and both sexes equally. Many other sets of parameter values have been investigated in computer experiments, but to simplify the presentation a decision was made to focus attention on the above values for the illustrative examples presented in this section.

The type of natural selection studied was that characterized by differential reproduction of mating types, a type not usually covered in books on population genetics. In Run 1, the 3×3 matrix of $\mu - -values$ was chosen as

$$\begin{bmatrix} 35 & 35 & 30 \\ 35 & 35 & 30 \\ 30 & 30 & 30 \end{bmatrix}, \tag{27}$$

indicating that all matings containing genotype aa were at a reproductive disadvantage. In Run 2, this matrix was chosen as

$$\begin{bmatrix} 10 & 10 & 8.57 \\ 10 & 10 & 8.57 \\ 8.57 & 8.57 & 8.67 \end{bmatrix}. \tag{28}$$

The ratio of the two distinct values in these matrices are approximately the same so that the two runs differed only in the absolute expected numbers of offspring contributed to the population per generation.

In each computer run, thirty iterates of the non-linear transformation were studied, representing thirty generations; fifty i.i.d Monte Carlo iterations were also made for each set of parameter values. Experimental evidence not reported here suggests that if the computer runs were extended to, say, 50 generations, the impressions of the behavior of the process as presented in the following graphs would not have changed substantially. That is, whenever a stationary types of behavior were suggested, these trends would have continued; similarly, increasing or decreasing trends would have continued.

Each Monte Carlo sample of size 50 was analyzed statistically as follows: Let $R(t)$ be any random function under consideration and let $R_i(t)$ $i = 1, 2, \cdots, 50$ be fifty Monte Carlo realizations of this random function in generation $t = 1, 2, \cdots, 30$. Then, the two extreme value statistics

$$Y_{50}(t) = \max \{ R_i(t) \mid i = 1, 2, \cdots, 50 \} \tag{29}$$

and

$$Y_1(t) = \min \{ R_i(t) \mid i = 1, 2, \cdots, 50 \}, \tag{30}$$

were computed at each generation $t = 0, 1, 2, \cdots, 30$. Similarly, the sample mean

$$\overline{Y}(t) = \frac{1}{50} \sum_{i=1}^{50} R_i(t) \tag{31}$$

was computed for each generation $t = 0, 1, 2, \cdots, 30$. Let $\widehat{R}(t)$ be the estimate of the random function $R(t)$ computed from the embedded non-linear deterministic model for each generation $t = 0, 1, 2, \cdots, 30$. Observe that the mean in (31) is an unbiased estimator of $E[R(t)]$, the unconditional expectation of R(t), which is usually non-computable in non-linear models. The strategy used in analyzing the computer experiments was to graph the functions just described simultaneously and, among other things, judge in what sense $\widehat{R}(t)$ is a measure of central tendency. Further technical details on the rationale underlying the method of statistical analysis just described may be found in [15] and the references contained therein. The programming language used in the computer implementation of the model was the DOS version of APL*PLUS II for a 486 Intel chip running at fifty mega-Hertz.

Figure 1 contains graphs of the extreme value statistics, the sample mean, which statistically summarize 50 Monte Carlo runs, as well as the numerical solution of the embedded non-linear difference equations for total population size plotted as functions of t measured in generations. As can be seen from this graph, total population size oscillated rather wildly among values in the hundreds to nearly 6,000, suggesting the model is chaotic at the parameter settings for Run 1. For this run, the solution of the embedded deterministic model, which is labeled DETERM in the legend, was not always a good measure of central tendency for the sample functions of the process; for at some time points, the sample mean was considerably above the point for the deterministic model and at some time points their positions were reversed. Moreover, at some time points, the deterministic model was close to the maximum of the fifty sample functions and at other points it was close the minimum. If the high level of productivity of offspring as represented by the expectations in (24) actually characterized a real population, then there would quite large and unpredictable fluctuations in total population size from generation to generation as indicated by the extreme value statistics. Had attention been confined solely to the embedded deterministic model, then these large fluctuations revealed by the stochastic model would have gone undetected.

When confronted with rather large fluctuations in total population size from generation to generation, a question that naturally arises is: what changes are occurring in the genetic composition of the population as described by the genotypic frequencies? Because at the parameter settings for Run 1 there is no differential viability between sexes, it suffices to confine attention to the female population. Figure 2 contains graphs of the genotypic frequencies for genotypes AA, Aa, and aa. As expected, because all matings containing genotype aa produce fewer offspring on the average, this genotype gradually decreased in frequency as can be seem for the MIN, MEAN, MAX, and DETERM graphs in Figure 2. However, the fluctuations, due to finite population size, in the frequencies of this genotype among Monte Carlo runs were considerable. For example, at generation 30, the minimum frequency for 50 Monte Carlo runs of this genotype was approximately 0.02, the mean was 0.12, the max was 0.26 and the frequency predicted by the deterministic

Figure 1: The MIN, MEAN,and MAX of Fifty Monte Carlo
Iterations and the Nummerical Solution of the Embedded
Deterministic Model, DETERM, for Total Population Size
in Run 1.

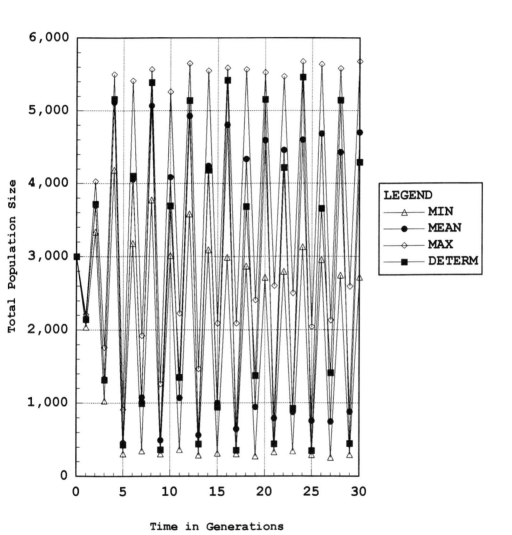

Figure 2: The MIN, MEAN, and MAX of Fifty Monte Carlo
Iterations and the Numerical Solution of the Embedded
Deterministic Model, DETERM, for the Genotypic Frequencies
in Run 1.

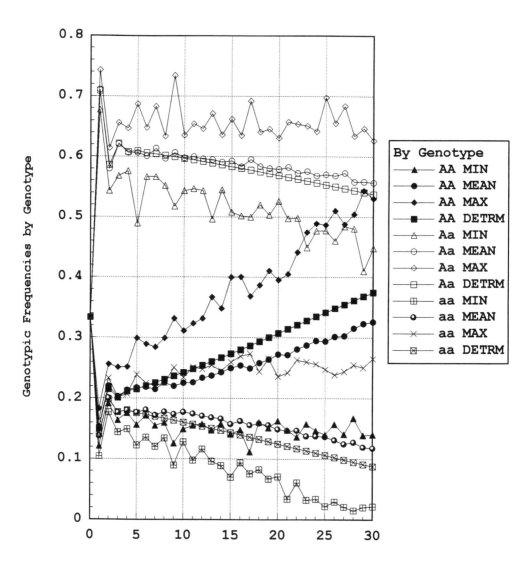

Figure 3: The MIN, MEAN,and MAX of Fifty Monte Carlo
Iterations and the Numerical Solution of the Embedded
Deterministic Model, DETERM, for Total Population Size
in Run 2.

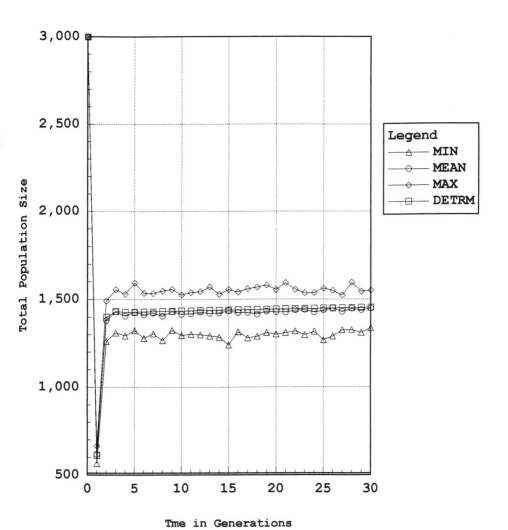

Figure 4: The MIN, MEAN, and MAX of Fifty Monte Carlo
Iterations and the Numerical Solution of the Embedded
Deterministic Model, DETERM, of the Genotypic Numbers
in Run 2.

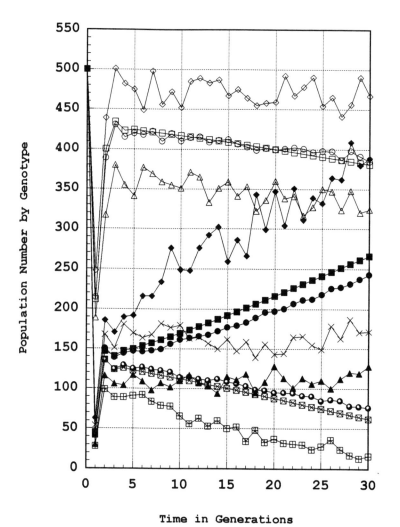
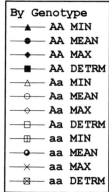

Time in Generations

model was 0.09, indicating that there could be considerable drift in genotypic frequencies due to fluctuations in population size that would not be detected by the deterministic model. Like those for the genotype aa, the frequencies of genotypes AA and Aa, as plotted in Figure 2, also exhibited considerable variations among the 50 Monte Carlo iterations.

A stereotypical view of deterministic models of population dynamics was that somehow their solutions represented essential central tendencies in the evolution of a population and that any deviation from these tendencies, due to stochastic effects, would be small, particularly in large populations. The results from Run 2 suggest that this view is only partially true. Presented in Figure 3 is the graphical summary of Run 2 with respect to total population size, a run in which each mating type on average produced fewer offspring, see (28). As can be seen from Figure 3, according to the deterministic model, total population size leveled off at a little less than 1,500 individuals, the mean of 50 Monte Carlo iterations closely followed that predicted by the embedded deterministic model, and, in comparison with Run 1, the extreme value statistics remained quite close to the mean, which in agreement with the stereotypical view. As one would expect, however, even for total population sizes as large as 1,500 but with several genotypes, the number of individuals of a each genotype could exhibit significant stochastic fluctuations. Figure 4 contains a summary of Run 2 with respect to the numbers of each of the three genotypes in the female population. Among the roughly 750 members of females in a steady state population, the number of individuals of genotype AA in generation 30 ranged from a minimum of 128 to a maximum of 389 in 50 Monte Carlo iterations; similar fluctuations, although smaller in magnitude, occurred for the other two genotypes Aa and aa. These results suggest that in populations whose size is restricted by the carrying capacity of the environment, there can be considerable stochastic fluctuations among the numbers of each of the genotypes and thus a drift in the genetic composition of the population from generation to generation.

7 Acknowledgement

It is a pleasure to acknowledge the help of Professor Edward Pollak, Department of Statistics, Iowa State University, who read the manuscript and offered many suggestions for improvement.

References

[1] Asmussen, S. and Hering, H. (1983) *Branching Processes*. Birkhauser,Boston.

[2] Mode, C. J. (1972) A Bisexual Multi-Type Branching Process and Applications in Population Genetics. *Bull. Math. Biophys.* 34,13-31.

[3] Mardia,K.V. (1976) *Families of Bivariate Distributions*. Griffin's Statistical Monographs, Hafner, Darien, Conn.

[4] Whitt, W. (1976) Bivariate Distributions with Given Marginals. *Ann.Statist.* 4,1280-1289.

[5] Johnson, N. L. and Kotz, S. (1972) *Distributions in Statistics - Continuous Multivariate Distributions.* Wiley, New York.

[6] Fristrom, J. W. and Spieth, P. T. (1980) *Principles of Genetics.* Chiron Press, New York and Concord.

[7] Crow, J. F. and Kimura, M. (1970) *An Introduction to Population Genetics Theory.* Harper and Row, New York.

[8] Cavalli-Sforza, L. L. and Bodmer, W. F. (1971) *The Genetics of Human Populations.* W. H. Freeman and Company, San Francisco.

[9] Harris, T. E. (1963) *The Theory of Branching Processes.* Springer-Verlag, Berlin.

[10] Mode, C. J. (1971) *Multitype Branching Processes - - Theory and Applications.* American Elsevier, New York.

[11] Athreya, K. B. and Ney, P. (1972) *Branching Processes.* Springer-Verlag, Berlin.

[12] Jagers, P. (1975) *Branching Processes with Biological Applications.* John Wiley and Sons, London.

[13] Karlin, S. and Taylor, H. M. (1975) *A First Course in Stochastic Processes - Second Edition.* Academic Press, Boston.

[14] Gulick, D. (1992) *Encounters With Chaos.* McGraw-Hill, New York.

[15] Mode, C. J. , Gollwitzer, H. E. , Salsburg, M. A. , and Sleeman. C. K. (1989) A Methodological Study of a Non-Linear Stochastic Model of an AIDS Epidemic with Recruitment. *IMA J. Math. App. Med. and Biol.* 6,179-203.

MULTITYPE CONTACT BRANCHING PROCESSES

J. RADCLIFFE AND L. RASS*

Abstract

Connections exist between n-type contact branching processes and deterministic models for spatial epidemics. Let the position of the furthest individual in a contact branching process from position 0 at time t in a given direction be denoted by $U(t)$. Define $y_i(s,t) = P[U(t) > s|$ one type i individual at position 0 at time $t = 0]$. This paper discusses how the methodology developed for considering the asymptotic speed of propagation of infection in n-type spatial epidemics can be modified to look at the behaviour of $y_i(s,t)$. This leads in certain cases to a proof of the result that $U(t)/t$ converges in probability to c_0, the minimum speed for which wave solutions exist in a particular system of equations. The application of an approximate saddle point method to more general contact branching processes is also discussed.

Multitype contact branching processes; spatial epidemics; asymptotic speed of propagation.

1. Introduction.

This paper discusses how methodology developed for the analysis of the behaviour of spatial models for the spread of a deterministic epidemic can be used in a stochastic context. These methods provide an analysis of certain contact branching processes.

The fundamental process that we wish to consider is a contact birth process in R^N. A simple contact birth process consists of only one type of individual. It commences at time $t = 0$ with a single individual at position $\mathbf{0}$. At a random time T which has a probability density function $\lambda e^{-\lambda t}, 0 \le t < \infty$, each individual gives rise to a further individual at a position $\mathbf{S}$ from the parent, where $\mathbf{S}$ is a vector of random variables. Let $U(t)$ be the position of the furthest individual from position $\mathbf{0}$ at time t in a given direction. We look at the behaviour of $y(s,t) = P[U(t) > s]$.

An n-type version of this process can also be formulated. The time until a birth event occurs still has an exponential distribution but its parameter depends upon the type of individual. When such an event occurs to a type i individual, then there is an associated probability that it will give rise to a type j individual. In this n-type process let $y_i(s,t) = P[U(t) > s|$ one type i individual at $s = 0$ at time $t = 0]$. Note that $U(t)$ can now either be the position of furthest spread of any individual at time t, or the position of furthest spread of a specific type of individual at time t. Then $y_i(s,t), i = 1, ..., n$ satisfy equations similar to those for the spatial spread of infection in an n-type $S \to I$ epidemic. For a single type process this was pointed out by Mollison

---

*School of Mathematical Sciences, Queen Mary and Westfield College, University of London, Mile End Road, London E14NS.

(1977,1978). A similar connection between the equation for the spread of a gene in a population process and the point of furthest spread in a branching diffusion process was first observed by McKean (1975). Contact birth-death and more general Markovian branching processes are also considered. The contact birth-death process links with the spatial $S \to I \to S$ epidemic.

For the contact birth, birth-death and more general Markovian branching processes, one can consider the approximate equations valid in the forward tail. In the forward region the approximate equations are identical with the approximate equations holding in the forward region of $S \to I$ and $S \to I \to S$ spatial epidemic models with suitably chosen parameters. The saddle point method can then be used to indicate the speed of first spread of $y_i(s,t)$ for these processes.

Rigorous analytic methods exist to examine the speed of propagation of infection for a generalisation of the $S \to I$ epidemic which allows for the possibility of varying infectivity or for removals and/or a latent period. In this paper we discuss how these methods can be adapted to give a rigorous derivation of the speed of propagation of $y_i(s,t)$ for the contact birth process. In the non reducible case this speed is c_0, which is the corresponding speed of propagation of infection and least velocity for which wave solutions exist in the $S \to I$ epidemic. The result we can obtain for the contact birth process starting with one type i individual, with contact distributions which are symmetric and are exponentially dominated in the tail, is that

$$\lim_{t \to \infty} P\left(\frac{U(t)}{t} > c\right) = \begin{cases} 0 & \text{for } c > c_0 \\ 1 & \text{for } c < c_0. \end{cases}$$

Hence $U(t)/t$ converges in probability to c_0. When the contact distributions are not all exponentially dominated in the tail, then $\lim_{t \to \infty} P\left(t^{-1}U(t) > c\right) = 1$ for all c.

The single type contact birth process has been investigated using probabilistic methods. The almost sure convergence of $U(t)/t$ was proved by Mollison (1977). The fact that it converged to c_0 was proved by Biggins (1978).

Results can also be obtained for the reducible case. In addition the exact results may be proved in a more general setting. This allows these results to also cover a simple form of n-type Markovian branching process in which there are no deaths. Rigorous analytic methods do not yet exist for the speed of propagation of the $S \to I \to S$ epidemic, so that the exact methods from the deterministic theory cannot at present be used for contact processes involving deaths.

2. Contact branching processes.

This section sets up the models which will be considered in this paper. These comprise the n-type versions of the contact birth process, the contact birth-death process and a particular specification of a contact branching process.

The contact birth process.

An n-type contact birth process in R^N commences with a single type i individual at position s=0 at time $t = 0$. The probability that a type i individual gives birth to an offspring in a time interval $(t, t + \delta t)$ is $\alpha_i \delta t + o(\delta t)$. The probability that this offspring is of type j is q_{ij}, where $\sum_j q_{ij} = 1$. The position of such a new individual relative to the parent has contact distribution $p_{ij}^*(s)$. A single type contact birth process was considered by Daniels (1977).

Consider the spread in a given direction. The projection of a contact branching process in a given direction is a contact branching process. Let $p_{ij}(s)$ be the marginal contact distribution in that direction Let U(t) be the position of the furthest individual from 0 at time t in the given direction, and let $y_i(s,t) = P[U(t) > s|$ one type i individual at position 0 at time $t = 0]$.

The time T until the first birth has probability density function $\alpha_i e^{-\alpha_i t}$, $0 \leq t < \infty$. Let $\lambda_{ij} = \alpha_i q_{ij}$. Note that $\alpha_i = \sum_{j=1}^n \lambda_{ij}$. Then $y_i(s,t) \equiv 1$ for $s < 0$ and $t \geq 0$. For $s \geq 0$, the usual conditional arguments lead to the following equations for $y_i(s,t)$:

$$(1 - y_i(s,t)) = P(T > t)(1 - y_i(s,0))$$
$$+ \int_0^t \alpha_i e^{-\alpha_i \tau}(1 - y_i(s,t-\tau)) \sum_{j=1}^n q_{ij} \int_{-\infty}^\infty (1 - y_j(s-r,t-\tau))p_{ij}(r)drd\tau. \tag{2.1}$$

Thus

$$(1 - y_i(s,t))e^{\alpha_i t} = (1 - y_i(s,0))$$
$$+ \int_0^t \alpha_i e^{\alpha_i \theta}(1 - y_i(s,\theta)) \sum_{j=1}^n \frac{\lambda_{ij}}{\alpha_i} \int_{-\infty}^\infty (1 - y_j(s-r,\theta))p_{ij}(r)drd\theta. \tag{2.2}$$

Differentiating with respect to t, the following equations are obtained for $s \geq 0$:

$$\frac{\partial y_i(s,t)}{\partial t} = (1 - y_i(s,t)) \sum_{j=1}^n \lambda_{ij} \int_{-\infty}^\infty y_j(s-r,t)p_{ij}(r)dr, \quad (i = 1,...,n). \tag{2.3}$$

Thus

$$\frac{\partial y_i(s,t)}{\partial t} = (1 - y_i(s,t)) \sum_{j=1}^n \lambda_{ij} \left[\int_{-\infty}^s y_j(s-r,t)p_{ij}(r)dr + \int_s^\infty p_{ij}(r)dr \right], \quad (i = 1,...,n). \tag{2.4}$$

Let

$$w_i(s,t) = \begin{cases} -\log(1 - y_i(s,t)) & \text{for } s \geq 0 \\ 0 & \text{for } s < 0. \end{cases}$$

We then obtain the equations

$$w_i(s,t) = \sum_{j=1}^n \lambda_{ij} \int_0^t \int_{-\infty}^\infty (1 - e^{-w_j(s-r,\tau)})p_{ij}(r)drd\tau + H_i(s,t), \tag{2.5}$$

for $s \geq 0$, where $H_i(s,t) = t \sum_{j=1}^n \lambda_{ij} \int_{-\infty}^s p_{ij}(r)dr$ for $s \geq 0$.

Note that if we consider the furthest spread of a specific type k, rather than any type, then equations may be set up in a similar manner. However, $y_i(s,0) = 0$ for all s if $i \neq k$; and $y_k(s,t) \equiv 1$ for $s < 0$ and $y_k(s,0) = 0$ for $s > 0$. We also define $w_i(s,t)$ somewhat differently. For $i \neq k$ we define $w_i(s,t) = -\log(1 - y_i(s,t))$ for all s, but

$$w_k(s,t) = \begin{cases} -\log(1 - y_k(s,t)) & \text{for } s \geq 0 \\ 0 & \text{for } s < 0. \end{cases}$$

Equations (2.3) and hence equations (2.5) are then obtained for all s if $i \neq k$ but for $s \geq 0$ if $i = k$, with an adjusted $H_i(s,t)$, namely $H_i(s,t) \equiv 0$ for all s if $i \neq k$ and $H_k(s,t) = t\sum_{j=1}^{n} \lambda_{kj} \int_{-\infty}^{s} p_{kj}(r)dr$ for $s \geq 0$.

In the forward tail where $(1 - y_i(s,t)) \approx 1$ equations (2.3), or its equivalent if we look at the spread of the kth type, can be approximated by the linear equations

$$\frac{\partial y_i(s,t)}{\partial t} = \sum_{j=1}^{n} \lambda_{ij} \int_{-\infty}^{\infty} y_j(s-r,t)p_{ij}(r)dr, \quad (i = 1,...,n). \tag{2.6}$$

The contact birth-death process.

The contact birth-death process is an extension of the above process where individuals can also die. The probability that an individual of type i dies in $(t, t + \delta t)$ is $\mu_i \delta t + o(\delta t)$. The time t until the first birth or death has probability density function $(\alpha_i + \mu_i)e^{-(\alpha_i + \mu_i)t}$, $0 \leq t < \infty$. Using a similar argument to that for the contact birth process, regardless of whether we consider the first spread of any type or of a specific type k, the following equations for $y_i(s,t)$ are obtained:

$$\frac{\partial y_i(s,t)}{\partial t} = (1 - y_i(s,t))\sum_{j=1}^{n} \lambda_{ij} \int_{-\infty}^{\infty} y_j(s-r,t)p_{ij}(r)dr - \mu_i y_i(s,t), \quad (i = 1,...,n). \tag{2.7}$$

Note that the equations hold for all s. If we consider the position of furthest spread of any type then

$$y_i(s,0) = \begin{cases} 1 & \text{for } s < 0. \\ 0 & \text{for } s \geq 0. \end{cases}$$

For the furthest spread of type k, $y_i(s,0) = 0$ for $i \neq k$ and

$$y_k(s,0) = \begin{cases} 1 & \text{for } s < 0. \\ 0 & \text{for } s \geq 0. \end{cases}$$

The approximate equations which hold in the forward tail are

$$\frac{\partial y_i(s,t)}{\partial t} = \sum_{j=1}^{n} \lambda_{ij} \int_{-\infty}^{\infty} p_{ij}(r)y_j(s-r,t)dr - \mu_i y_i(s,t), \quad (i = 1,...,n). \tag{2.8}$$

A contact branching process.

The generalisation to a contact branching process may be carried out in a number of ways. Results for a single type branching process where individuals die when they give birth have been obtained using probabilistic methods by Uchiyama (1982). See also Biggins (1978). When a birth event occurs the number of offspring of different types will have a distribution. The positions of these offspring may all be the same, be independent but depend on the type, or have a more general joint distribution. Each of these models will be specified by equations analogous to equations (2.3) and (2.7). The approximate method is very powerful in that it can be used to treat quite general branching processes and is applicable to all these models. The exact method requires quite delicate analysis and can, at present, only be used for the contact birth process and a special case of contact branching process without deaths. We now formulate this contact branching process, but allow for deaths. This includes the contact birth-death process as a special case.

The times until a birth event or a death of an individual have probability densities $\alpha_i e^{-\alpha_i t}$, $0 \le t < \infty$, and $\mu_i e^{-\mu_i t}$, $0 \le t < \infty$ respectively. The probability that the type i individual gives rise to m offspring is $f_i(m)$, all offspring being restricted to being of the same type. The probability that they are of type j is q_{ij}. The density function of the distance r in a specific direction at which they are all situated relative to the individual giving birth is $p_{ij}(r)$. The equations obtained for the $y_i(s,t)$ are

$$
\frac{\partial y_i(s,t)}{\partial t} = (1 - y_i(s,t)) \sum_{j=1}^{n} \alpha_i q_{ij} \int_{-\infty}^{\infty} (1 - \pi_i(1 - y_j(s-r,t))) p_{ij}(r) dr
$$
$$
- \mu_i y_i(s,t), \quad (i = 1, ..., n),
$$
(2.9)

where $\pi_i(\theta) = \sum_{m=1}^{\infty} f_i(m) \theta^m$.

Consider the speed of spread in the forward tail. In the tail $1 - y_i(s,t) \approx 1$, so we obtain the approximate linear equations

$$
\frac{\partial y_i(s,t)}{\partial t} = \sum_{j=1}^{n} \alpha_i q_{ij} \pi_i'(1) \int_{-\infty}^{\infty} y_j(s-r,t) p_{ij}(r) dr - \mu_i y_i(s,t), \quad (i = 1, ..., n). \quad (2.10)
$$

If individuals cannot die, so that $\mu_i = 0$, $i = 1, ...n$, and we define $w_i(s,t)$ in terms of $y_i(s,t)$ as for the birth process, the equations analogous to equations (2.5) are

$$
w_i(s,t) = \sum_{j=1}^{n} \lambda_{ij} \int_0^t \int_{-\infty}^{\infty} g_i(w_j(s-r,\tau)) p_{ij}(r) dr d\tau + H_i(s,t), \quad (i = 1, ..., n), \quad (2.11)
$$

where $\lambda_{ij} = \alpha_i q_{ij} \pi_i'(1)$ and $g_i(\theta) = (1 - \pi_i(e^{-\theta}))/\pi_i'(1)$. The conditions for $H_i(s,t)$ are the same as for the contact birth process, as are the restrictions on s, i for which equations (2.11) hold. Equations (2.5) are a special case of equations (2.11) with $\pi(\theta) = \theta$ for $i = 1, ..., n$. It is this process which we can treat using the exact method. In this treatment we require $g_i(0) = 0$ and $g_i'(0) = 1$. It is interesting to note that when considering one type spatial models in genetics, equations similar to equations (2.11) arise with $n = 1$ and a different function $g_i(\theta)$. (see Diekman and Kaper (1978) and Lui (1982a,1982b)).

3. The connection with deterministic spatial epidemics.

This section gives the connections between the contact processes of Section 2 and deterministic spatial epidemics. The strongest connections are between the contact birth process and the $S \to I$ epidemic, and the contact birth-death process and the $S \to I \to S$ epidemic. In each case the models are described by the identical equations, apart from a minor modification of the parameters. However, the functions involved satisfy different conditions, so that results for the contact processes do not follow immediately from those for the corresponding epidemic models. For more general models the connection with the epidemic models is via the approximate equations which are valid in the forward tails.

The contact birth and S $\to$ I epidemic connection.

Consider an epidemic taking place amongst n types in R, the ith type having uniform density σ_i. The rate of infection of susceptible individuals of type i by infectious individuals of type j is λ_{ij}. Let $y_i(s,t)$ be the proportion of infectious individuals at position s at time t. Suppose there is some infection present at time $t = 0$, so that $y_i(s,0) \neq 0$ for some s and i. Then the $y_i(s,t)$ satisfy the equations

$$\frac{\partial y_i(s,t)}{\partial t} = (1 - y_i(s,t)) \sum_{j=1}^{n} \lambda_{ij} \sigma_j \int_{-\infty}^{\infty} y_j(s-r,t) p_{ij}(r) dr, \quad (i = 1, ..., n). \quad (3.1)$$

Let $w_i(s,t) = -\log(1 - y_i(s,t))$. We then obtain the equations

$$w_i(s,t) = \sum_{j=1}^{n} \sigma_j \lambda_{ij} \int_0^t \int_{-\infty}^{\infty} (1 - e^{-w_j(s-r,\tau)}) p_{ij}(r) dr d\tau + H_i(s,t), \quad (3.2)$$

where $H_i(s,t) = w_i(s,0) = -\log(1 - y_i(s,0))$.

An alternative formulation is to initiate infection at time $t = 0$ amongst the populations by the introduction of infection from outside. This leads to the equations

$$\frac{\partial y_i(s,t)}{\partial t} = (1 - y_i(s,t)) \left(\sum_{j=1}^{n} \lambda_{ij} \sigma_j \int_{-\infty}^{\infty} y_j(s-r,t) p_{ij}(r) dr + h_i(s,t) \right), \quad (i = 1, ..., n),$$

$$(3.3)$$

where $h_i(s,t)$ is a term giving the effect due to the individuals from outside. This leads again to equations (3.2), with $H_i(s,t) = \int_0^\infty h_i(s,\tau) d\tau$ and $w_i(s,0) \equiv 0$.

In either case equations (3.2) for the $S \to I$ epidemic are identical to equations (2.5) for the contact birth process, provided that we replace $\lambda_{ij}\sigma_j$ for the epidemic by the parameter λ_{ij} for contact birth process. Note, however, that equations (3.2) hold for all s, unlike equations (2.5) for the contact birth process. When studying such an epidemic the infection from outside was taken to be in a bounded region of R and to initiate infection only so that $H_i(s,t)$ is uniformly bounded over all s and t. For the contact birth process $H_i(s,t)$ is not uniformly bounded. There is, however some simplicity in the contact birth process. Each $H_i(s,t)$ is monotone in s, as also is each $w_i(s,t)$. This is not true for the $S \to I$ epidemic.

The contact birth-death and S → I → S epidemic connection.

An $S \to I \to S$ epidemic allows individuals to return to the susceptible state. The rate at which a type i infective returns to the susceptible state is μ_i. We consider the case where some infection is present at time $t = 0$, i.e. $y_i(s,0) \neq 0$ for some s and i. The equations are

$$\frac{\partial y_i(s,t)}{\partial t} = (1 - y_i(s,t)) \sum_{j=1}^{n} \lambda_{ij}\sigma_j \int_{-\infty}^{\infty} y_j(s-r,t)p_{ij}(r)dr - \mu_i y_i(s,t), \quad (i = 1,...,n).$$

$$(3.4)$$

If we replace $\lambda_{ij}\sigma_j$ by λ_{ij}, these equations are identical to equations (2.7) for the contact birth-death process. However, the initial values $y_i(s,0)$ differ. For the contact birth-death process we again have monotonicity in s.

The spread of the tail of the distribution for contact branching processes.

For the more general process described in Section 2 there is no longer such a simple connection. However, if we consider how fast the forward tail of the distribution of U(t) spreads, we can approximate the equations as discussed in Section 2. Note that the various contact branching processes lead to the same approximate equations. These are precisely the same equations that are obtained for the $S \to I \to S$ epidemic by taking $(1 - y_i(s,t)) \approx 1$ in the tail, provided that we replace $\lambda_{ij}\sigma_j$ for the epidemic by the parameter $\alpha_i q_{ij} \pi_i'(1)$ for the contact branching process.

The approximate equations for a contact branching process without deaths correspond to the approximate equations for the $S \to I$ epidemic.

4. The exact method.

In Section 3, we showed that both the contact birth process and the $S \to I$ epidemic lead to the same equations (2.5) for $w_i(s,t)$, but with adjusted parameters and a restriction on the the validity of the equations for the contact birth process. Also the functions $H_i(s,t)$ differ. Exact results have been proved for a generalisation of an $S \to I$ epidemic to allow for non-constant infectivity. This generalisation also covers models with removals and/or a latent period. These models and their analyses are described in a series of papers (Radcliffe and Rass (1983,1984a,1986,1991)). For the single type case a model with varying infectivity has been considered by Diekmann (1978). Note, however, that the conditions on this model were such that it did not cover the $S \to I$ epidemic. See also Thieme (1979a,1979b). The proofs require $H_i(s,t)$ to be uniformly bounded and (for an epidemic in R) the $p_{ij}(r)$ to be symmetric about zero. The contact birth process has restrictions on the values of s for each i for which equations (3.2) are valid. The restrictions depend upon whether we consider the spread of any type or of a specific type k. In addition the $H_i(s,t)$ are not uniformly bounded. Hence the results for the contact birth process are not immediate. The proofs for the deterministic epidemic model can be adapted to cover not only the contact birth process, but also the particular contact branching process described in section 2. The analysis is therefore based on equations (2.11). The methodology used to establish the speed of propagation is quite complex. We therefore restrict ourselves to a summary of the main results. As for the epidemic models, the $p_{ij}(r)$ are restricted to be symmetric. We first consider a non-reducible contact birth process. In such a process every type can, perhaps only several generations later, give rise to descendants of any type. This implies that $\Lambda = (\lambda_{ij})$ is a non-reducible matrix.

Consider first what is meant by the asymptotic speed of propagation of the forward tail of the distribution of $U(t)$, where initially there is one type i individual at position zero. Here $y_i(s,t) = P(U(t) > s)$. A natural definition, since $y_i(s,t)$ is monotone in s, is to consider a small $\eta > 0$ and to find $s(t)$ such that $y_i(s(t),t) = \eta$. The speed of propagation would be $\lim_{t\to\infty} s(t)/t$. This we term Definition 1.

Aronson and Weinberger (1975,1978) defined the speed of spread for $y_i(s,t)$, for the deterministic epidemic, in a somewhat different manner. We refer to this as Definition 2. The speed of propagation is said to be c^* if the following two conditions hold:

(i) For any $c > c^*$, $\lim_{t\to\infty} \sup\{y_i(s,t) : |s| \geq ct\} = 0$

(ii) For each c such that $0 < c < c^*$ there exists an $\varepsilon > 0$ and $T > 0$ such that $\min\{y_i(s,t) : |s| \leq ct\} \geq \varepsilon$ for all $t \geq T$.

For the contact branching process we consider the speed of spread in the positive direction only. Also $y_i(s,t)$ is monotone in s for each t. The conditions above therefore become:

(i) For any $c > c^*$, $\lim_{t\to\infty} P\left(t^{-1}U(t) > c\right) = \lim_{t\to\infty} y_i(ct,t) = 0$.

(ii) For each c such that $0 < c < c^*$ there exists an $\varepsilon > 0$ and $T > 0$ such that $P\left(t^{-1}U(t) > c\right) = y_i(ct,t) \geq \varepsilon$ for all $t \geq T$.

The exact methods of our paper (1986) for the deterministic epidemic first establish the asymptotic speed of propagation using Definition 2. The results pertaining to part (ii) are then strengthened and used to prove the pandemic theorem, which gives a lower bound on the proportions eventually suffering the epidemic. The method, when applied to the contact birth process, enables us to show that, for a specified c^*, $\lim_{t\to\infty} y_i(ct,t) = 0$ for any $c > c^*$ and $\lim_{t\to\infty} y_i(ct,t) = 1$ for any $c < c^*$. Hence

$$\lim_{t\to\infty} P\left(\frac{U(t)}{t} > c\right) = \begin{cases} 0 & \text{for } c > c^* \\ 1 & \text{for } c < c^*. \end{cases}$$

Then $U(t)/t$ converges in probability to c^*

Observe that when we prove these results they also imply that the asymptotic speed of propagation as given by definition 1 is also c^*. For c^* finite, necessarily $s(t)/t$ tends to a limit which is c^*. When c^* is infinite, then $s(t)/t$ tends to infinity as t tends to infinity.

When each of the $p_{ij}(r)$ are exponentially dominated in the tail we define $P_{ij}(\theta) = \int_{-\infty}^{\infty} e^{\theta r} p_{ij}(r)dr$. Let Δ_{ij} be the abscissa of convergence of $P_{ij}(\theta)$ in the positive half of the complex plane, and let $\Delta_V = \min_{ij} \Delta_{ij}$. Define $c_0 = \inf_{\theta \in (0, \Delta_V)} \rho((\lambda_{ij} P_{ij}(\theta)))/\theta$, where $\rho(A)$ is the Perron-Frobenius root of the non-negative non-reducible matrix A. We can prove the following theorems.

Theorem 1

When the $p_{ij}(r)$ are all exponentially dominated in the tail there exists a monotone (in t) solution $w_i(s,t)$ to equations (2.11) which is unique. For any c such that there exists a $\theta \in (0, \Delta_V)$ with $c > \rho((\lambda_{ij} P_{ij}(\theta)))/\theta$, then $\lim_{t\to\infty} w_i(ct,t) = 0$ for $i = 1, ..., n$.

Corollary 1
For any $c > c_0$ $\lim_{t \to \infty} w_i(ct, t) = 0$ for $i = 1, ..., n$. Therefore when we start with one type i individual at position 0, $\lim_{t \to \infty} P\left(t^{-1}U(t) > c\right) = 0$ for $c > c_0$.

Theorem 2
If at least one $p_{ij}(r)$ is not exponentially dominated in the tail take any $c > 0$, otherwise take any c such that $0 < c < c_0$. Then for any positive constants B_i there exist T_i sufficiently large so that $w_i(ct, t) \geq B_i$ for all $t \geq T_i$.

Corollary 2
The following results hold for all c if at least one $p_{ij}(r)$ is not exponentially dominated in the tail and for $c < c_0$ if all the $p_{ij}(r)$ are exponentially dominated in the tail. For any $\epsilon_i > 0$ there exist T_i sufficiently large so that $y_i(ct, t) \geq 1 - \epsilon_i$ for all $t \geq T_i$. If we start with one type i individual at position 0 then $\lim_{t \to \infty} P\left(t^{-1}U(t) > c\right) = 1$.

Theorems 1 and 2 together show that when the $p_{ij}(r)$ are all exponentially dominated in the tail the asymptotic speed of propagation is $c_0 = \inf_{\theta \in (0, \Delta_V)} \rho((\lambda_{ij} P_{ij}(\theta)))/\theta$, and that $U(t)/t$ converges in probability to c_0. When at least one $p_{ij}(r)$ is not exponentially dominated in the tail the asymptotic speed of propagation is infinite. Hence $\lim_{t \to \infty} P\left(t^{-1}U(t) > c\right) = 1$ for all c.

It is interesting to note that for the contact birth process c_0 can be related to the speeds of possible wave solutions of the equations for the $S \to I$ epidemic. When all the $p_{ij}(r)$ are exponentially dominated in the tail, c_0 is the infimum of the wave speeds for which non-trivial wave solutions exist to equations (2.3). Except in an exceptional case, there is a unique wave solution modulo translation at speed c_0, (Radcliffe and Rass (1984a)).

5. The approximate method for contact branching processes.

A saddle-point method can be used on the approximate equations (2.10) valid in the forward tail of the distribution of U(t) for general contact branching processes with deaths. This method will give the asymptotic speed of first spread. It does not require the contact distributions to be symmetric about zero, but does require them to be exponentially dominated in the forward tail. The saddle-point method used for deterministic epidemics is described in our papers (Radcliffe and Rass (1984b, 1993)). See also Daniels (1975) Note that for the $S \to I \to S$ epidemic $y_i(s, t)$ is not monotone in s, for fixed t. Hence, given a small $\eta > 0$, we cannot uniquely define $s(t) : y_i(s, t) = \eta$ and hence find $\lim_{t \to \infty} s(t)/t$. The approach used therefore was to define $s(t) : \int_{s(t)}^{\infty} y_i(x, t)dx = \eta$ and to use the saddle point method to find $\lim_{t \to \infty} s(t)/t$.

For the contact branching proces $\int_s^{\infty} y_i(x, t)dx = \int_s^{\infty}(x - s)f_{U(t)}(x)dx$, where $f_{U(t)}(x)$ is the density function of $U(t)$. This approach would consider the speed at which the weighted tail of the distribution of $U(t)$ first moves out. Because of the monotonicity of $y_i(s, t)$ in s, we can in fact use the simpler approach for the contact branching processes i.e take $s(t) : y_i(s, t) = \eta$ and find $\lim_{t \to \infty} s(t)/t$. Let $\mathbf{L}(\theta, t)$ be the vector of Laplace transforms of the $y_i(s, t)$. From equations (2.10) we obtain

$$\frac{\partial \mathbf{L}(\theta, t)}{\partial t} = ((\lambda_{ij} P_{ij}(\theta)) - diag(\mu_1, ..., \mu_n)) \mathbf{L}(\theta, t),$$

where $\lambda_{ij} = \alpha_i q_{ij} \pi'_i(1)$. Hence

$$
\begin{aligned}
\mathbf{L}(\theta, t) &= \exp\left(((\lambda_{ij} P_{ij}(\theta)) - diag(\mu_1, ..., \mu_n))\mathbf{L}(\theta, 0)\right) \\
&= e^{(\mathbf{A}(\theta) - \mu \mathbf{I})t}\mathbf{L}(\theta, 0),
\end{aligned}
$$

with $\mathbf{A}(\theta) = (\lambda_{ij} P_{ij}(\theta)) + diag(\mu - \mu_1, ..., \mu - \mu_n)$, and $\mu = \max(\mu_1, ..., \mu_n)$
The transform $\{\mathbf{L}(\theta, t)\}_j$ is inverted, so that, for a suitable choice of $c(t)$,

$$
y_j(s, t) = \frac{1}{2\pi i} \int_{c(t) - i\infty}^{c(t) + i\infty} e^{-s\theta} \{\mathbf{L}(\theta, t)\}_j \, d\theta.
$$

For a given small $\eta > 0$, $s(t)$ is such that $y_j(s(t), t) = \eta$. Take $\theta = c(t)$ to be the saddle point of $Re(g(\theta))$, where $g(\theta) = (\rho(\mathbf{A}(\theta))t - \theta s(t))$. The speed of spread is found to be $\lim_{t \to \infty} s(t)/t = \max\left(0, \inf_{\theta \in (0, \Delta_V)} f(\theta)\right)$, where $f(\theta) = (\rho(\mathbf{A}(\theta)) - \mu)/\theta$. For details of the saddle point method see (Radcliffe and Rass (1984)). Note that when $\mu = 0$ these results are identical to those obtained in section 4 for the special type of contact branching processes without deaths, in the case when the $p_{ij}(r)$ symmetric about zero and exponentially dominated in the tail.

Results can also be obtained for the reducible case. There are some technical conditions under which the proofs can be obtained. For an account of the results using the approximate method for the reducible epidemic see Radcliffe and Rass (1993). The exact method may also be used to rigorise the results obtained by the approximate method for the contact birth process and the special contact branching process. Note that the exact method requires the contact distributions to be symmetric.

References

Aronson, D. G. & Weinberger, H. F. (1975), Nonlinear diffusion in population genetics, combustion, and nerve pulse propagation. Goldstein, J. A. (ed.) Partial differential equations and related topics, Lect. Notes Math., No 446. Springer, Berlin, Heidelberg, New York. 5-49.

Aronson, D. G. & Weinberger, H. F. (1978), Multidimensional nonlinear diffusion arising in population genetics. *Adv. Math.*, **30**, 33-76.

Biggins, J. D. (1978), The asymptotic shape of the branching random walk. *Adv. Appl. Prob.*, **10**, 62-84.

Daniels, H. E. (1975), The deterministic spread of a simple epidemic. Perspectives in Probability and Statistics: Papers in Honour of M. S. Bartlett, Gani J. (ed.) Distributed for Applied Probability Trust by Academic Press, London. pp. 689-701.

Daniels, H. E. (1977), The advancing wave in a spatial birth process. *J. Appl. Prob.*, **14**, 689-701.

Diekmann, O. (1978), Run for your life. A note on the asymptotic speed of propagation of an epidemic. *J. Diff. Equations* **33**, 58-73.

Diekmann, O. & Kaper, H. G. (1978), On the bounded solutions of a nonlinear convolution equation. *Nonlin. Anal. Theory Appl.*, **2**, 721-737.

Lui, R. (1982a), A nonlinear integral operator arising from a model in population genetics, I. monotone initial data. *SIAM J. Math. Anal.*, **13** 913-937.

Lui, R. (1982b), A nonlinear integral operator arising from a model in population genetics, II. initial data with compact support. *SIAM J. Math. Anal.,* **13** 938-953.

McKean, H. P. (1975), Application of Brownian motion to the equation of Kolmogorov-Petrovskii-Piscunov. *Commun. Pure Appl. Maths.,* **28**, 323-331.

Mollison, D. M. (1977), Spatial contact models for ecological and epidemic spread. *J. R. Statist. Soc. B.,* **39**, 283-326.

Mollison, D. M. (1978), Markovian contact processes. *Adv. Appl. Prob.,* **10**, 85-108.

Radcliffe J. & Rass, L. (1983), Wave solutions for the deterministic non-reducible n-type epidemic. *J. Math. Biol.,* **17**, 45-66.

Radcliffe, J. & Rass, L. (1984a), The uniqueness of wave solutions for the deterministic non-reducible n-type epidemic. *J. Math. Biol.,* **19**, 303-308.

Radcliffe, J. & Rass, L. (1984b), Saddle-point approximations in n-type epidemics and contact birth processes. *Rocky Mountain J. Math.,* **14**, 599-617.

Radcliffe, J. & Rass, L. (1986), The asymptotic speed of propagation of the deterministic non-reducible n-type epidemic. *J. Math. Biol.,* **23**, 341-359.

Radcliffe, J. & Rass, L. (1991), The effect of reducibility on the deterministic spread of infection in a heterogeneous population. The Second International Conference on Mathematical Population Dynamics, Arino, O., Axelrod, D. E. and Kimmel M.,(ed.) Lecture Notes in Pure and Applied Mathematics, Vol 131, Marcel Dekker, pp. 93-114.

Radcliffe, J. & Rass, L. (1993), Reducible epidemics : choosing your saddle. *Rocky Mountain J. Math.,* **23**, 725-752.

Thieme, H. R. (1979a), Asymptotic estimates of the solutions of nonlinear integral equations and asymptotic speeds for the spread of populations. *J. Reine. Angew. Math.,* **306**, 94-121.

Thieme, H. R. (1979b), Density-dependent regulation of spatially distributed populations and their asymptotic speed of spread. *J. Math. Biol.,* **8**, 353-396.

Uchiyama, K. (1982), Spatial growth of a branching process of particles living in R^d. *Ann. Prob.,* **10**, 896-918.

Lecture Notes in Statistics

For information about Volumes 1 to 12
please contact Springer-Verlag